SpringerBriefs in Electrical and Computer Engineering

T0214338

More information about this series at http://www.springer.com/series/10059

Xu Chen · Jianwei Huang

Social Cognitive
Radio Networks

Xu Chen
University of Göttingen
Göttingen
Germany

Jianwei Huang
The Chinese University of Hong Kong
Shatin
Hong Kong SAR

ISSN 2191-8112   ISSN 2191-8120   (electronic)
SpringerBriefs in Electrical and Computer Engineering
ISBN 978-3-319-15214-1   ISBN 978-3-319-15215-8   (eBook)
DOI 10.1007/978-3-319-15215-8

Library of Congress Control Number: 2014960250

Springer Cham Heidelberg New York Dordrecht London

Printed on acid-free paper

Springer International Publishing AG Switzerland is part of Springer Science+Business Media
(www.springer.com)

Recommended by Sherman Shen.

Preface

Wireless spectrum is a scarce resource, and historically it has been divided into chunks and allocated to different government and commercial entities with long-term and exclusive licenses. This approach protects license users from harmful interferences from unauthorized users, but leaves little spectrum for emerging new services and leads to low spectrum utilizations in many spectrum bands. The way to turn spectrum drought into spectrum abundance is to allow dynamic and opportunistic spectrum sharing between primary licensed and secondary unlicensed users with different priorities. Such sharing is becoming technologically feasible due to the recent advances such as cognitive radio and small cell technologies, which allow multiple wireless devices to transmit concurrently in the same spectrum without significant mutual negative impacts.

As the spectrum opportunities are often dynamically changing over frequency, time, and space due to primary users' stochastic traffic, secondary users need to make intelligent spectrum access and sharing decisions. In this book, we propose a novel social cognitive radio networking framework—a transformational and innovative networking paradigm that promotes the nexus between social interactions and distributed spectrum sharing. By leveraging the wisdom of crowds, the secondary users can overcome various challenges due to incomplete network information and limited capability of individual secondary users. Building upon the social cognitive radio networking principle, we develop three socially inspired distributed spectrum sharing mechanisms: adaptive channel recommendation mechanism, imitative spectrum access mechanism, and evolutionarily stable spectrum access mechanism. Numerical results also demonstrate that the proposed socially inspired distributed spectrum sharing mechanisms can achieve superior networking performance.

The outline of this book is as follows. Chapter 1 overviews the related literature and discusses the motivations of social cognitive radio networking. Chapter 2 presents the adaptive channel recommendation mechanism, which is inspired by the recommendation system in the e-commerce industry for collaborative information filtering. Chapter 3 presents the imitative spectrum access mechanism, which leverages the common social phenomenon "imitation" to achieve efficient and fair distributed spectrum sharing. Chapter 4 presents the evolutionarily stable spectrum access mechanism, which is motivated by the evolution rule observed in many animal and human social interactions. Chapter 5 summarizes the main results in this book.

We would like to thank the series editor, Prof. Xuemin (Sherman) Shen from University of Waterloo, for encouraging us to prepare this monograph. We also want to thank members of the Network Communications and Economics Lab (NCEL) at the Chinese University of Hong Kong, for their supports during the past several years.

The work described in this book was supported by grants from the Research Grants Council of the Hong Kong Special Administrative Region, China (Project No. CUHK 412713 and CUHK 14202814). It is also partially supported by the funding from Alexander von Humboldt Foundation. Part of the results have appeared in our prior publications [1–3] and in the first authors Ph.D. dissertation [4].

Göttingen, Germany   Xu Chen
Hong Kong, China   Jianwei Huang

References

1. X. Chen, J. Huang, H. Li, Adaptive channel recommendation for opportunistic spectrum access. IEEE Trans. Mob. Comput. **12**(9), 1788–1800 (2013), Available: http://arxiv.org/pdf/1102.4728.pdf
2. X. Chen, J. Huang, Imitation-based social spectrum sharing. IEEE Trans. Mob. Comput. (2014). Available: http://arxiv.org/pdf/1405.2822v1.pdf
3. X. Chen, J. Huang, Evolutionarily stable spectrum access. IEEE Trans. Mob. Comput. **12** (7), 1281–1293 (2013). Available: http://arxiv.org/pdf/1204.2376v1.pdf
4. X. Chen, *Distributed spectrum sharing: a social and game theoretical approach* (The Chinese University of Hong Kong, Hong Kong, 2012). Ph.D. Dissertation.

Contents

Chapter 1
Overview

1.1 Spectrum Under-Utilization Issue

Global mobile traffic has been growing rapidly in the past several years [1]. Not only the average smartphone data usage tripled in 2011, but the non-smartphone wireless traffic also more than doubled in the same year. These sharp increases in mobile traffic are expected to continue in the foreseeable future [1]. In July 2011, Credit Suisse reported that wireless base stations in the United States were operating at 80 % of their maximum capacity during busy periods [2]. Compounding the issue of congested cellular networks is the wide use of social networking applications on mobile devices, where a viral social content can have a rapid increase in popularity in a short time (called a flash crowd [3]) and contributes to the significance increase of mobile data usage. This combination of exploding data demands and limited wireless resources poses a significant challenge for future wireless network design.

To address this challenge, regulatory agencies (e.g., FCC in U.S. and Ofcom in U.K.) around the world are actively working on the reformation of wireless spectrum access policies and regulations. Traditionally, wireless spectrum is regulated under the static and exclusive spectrum management policy, such that spectrum is allocated to spectrum licensees over large geographical areas for years or even decades [4]. A network operator (who is often a spectrum licensee) will use the licensed spectrum exclusively to serve his own primary licensed users. As a result, secondary unlicensed users cannot access the licensed bands under the static license arrangement. Since most spectrums have been licensed to different government and commercial entities, this will soon lead to the spectrum drought for many emerging new wireless services.

On the other hand, however, many existing licensed spectrum bands are not always efficiently utilized. According to [4], the temporal and spatial variations in the utilization of the licensed spectrum rang from 15 to 85 %, with a large portion of licensed spectrum being severely under-utilized. A field measurement by Shared Spectrum Cooperation shows that the overall average utilization of a wide range of different

© The Author(s) 2015

X. Chen and J. Huang, *Social Cognitive Radio Networks*,
SpringerBriefs in Electrical and Computer Engineering,
DOI 10.1007/978-3-319-15215-8_1

types of spectrum bands is lower than 20 % even in densely populated cities such as Chicago and New York City [5].

To address the spectrum under-utilization problem and support the growing wireless traffic demand, a novel dynamic spectrum sharing approach has been proposed [6]. Dynamic spectrum sharing enables unlicensed secondary wireless users equipped with cognitive radios to opportunistically share the spectrum with licensed primary users, in order to improve the spectrum utilization. A key challenge of dynamic spectrum sharing is how to achieve efficient spectrum sharing among secondary users in a distributed fashion. This is because that the spectrum opportunities for secondary users are often dynamically changing over frequency, time, and space due to stochastic traffic of primary users, and individual secondary users often have limited information of the entire network environment due to hardware constraints. Furthermore, if too many secondary users utilize the same vacant spectrum simultaneously, they would generate severe interferences to each other, leading to a poor system performance. Achieving an efficient distributed spectrum sharing thus requires that each secondary user has the ability to make intelligent decisions based on limited network information.

1.2 Social Cognitive Radio Networks

To overcome this challenge, a large body of literature has focused on investigating the *individual intelligence* of secondary users. For the individual intelligence, secondary users act with *full rationality* and share the spectrum through noncooperative competitions. Noncooperative game theory has been widely used to model the complex interactions among competitive secondary users and compute the best response based spectrum access strategy. To have full rationality, however, a secondary user typically needs to have a high computational power to collect and analyze the network information in order to predict other users' behaviors. This is often not feasible due to the limitations of today's mobile devices.

Along a different line, in this book we explore the *social intelligence* of secondary users for achieving an efficient distributed spectrum sharing. For the social intelligence, secondary users act with *bounded rationality* and share the spectrum through cooperative social interactions. The motivation for considering social intelligence is, by leveraging the wisdom of crowds, to overcome the challenges due to incomplete network information and limited capability of individual secondary users. In fact, the emergence of social intelligence has been observed in many social interactions of animals [7], and has been utilized for engineering algorithm design. For example, Kennedy and Eberhart designed the particle swarm optimization algorithm by simulating social movement behaviors in a bird flock [8]. Pham et al. developed the bees algorithm by mimicing the food foraging behaviors of honey bees [9]. The understanding of human social phenomenon also sheds new light into the design of more efficient engineering systems such as wireless communication networks. For example, the small-world phenomenon in social networks has been applied to design efficient decentralized routing strategy and topology control algorithms for ad hoc networks in [10, 11], respectively.

Building upon the principle of social intelligence, in this book we propose a novel social cognitive radio networking paradigm that promotes the nexus between social interactions and distributed spectrum sharing. Specifically, we develop three socially inspired distributed spectrum sharing mechanisms: (1) Inspired by the recommendation system in the e-commerce industry such as Amazon, we propose an *adaptive channel recommendation mechanism*, such that secondary users collaboratively recommend "good" channels to each other for achieving more informed spectrum access decisions; (2) By leveraging a common social phenomenon "imitation" in human and animal society, we devise an *imitative spectrum access mechanism*, such that secondary users imitate the spectrum access strategies of their elite neighbours to improve the networking performance; (3) Motivated by the evolution rule observed in many animal and human interactions, we propose an *evolutionarily stable spectrum access mechanism*, such that each secondary user takes a comparison strategy (i.e., compare its performance with the collective network performance) to evolve its spectrum access decision adaptively over time.

1.3 Related Research

For the individual intelligence, a common modeling approach is to consider selfish secondary users, and model their interactions as non-cooperative games. There is a vast literature along this line, and here we will briefly outline some representative ones. Nie and Comaniciu [12] designed a self-enforcing distributed spectrum access mechanism based on potential games. Niyato and Hossain [13] proposed a dynamic game approach for analyzing the competition among secondary users for spectrum access. Flegyhzi et al. [14] proposed a two-tier game framework for cognitive radio medium access control (MAC) mechanism design. Yang et al. [15] studied a price-based spectrum access mechanism for competitive secondary users. Li et al. [16] proposed a game theoretic framework to achieve incentive compatible multiband sharing among the secondary users. Chen and Huang [17, 18] developed a spatial spectrum access game framework to model the competitive spectrum access among the secondary users by taking the spatial reuse effect into account. Southwell et al. [19] studied the distributed QoS satisfaction for spectrum sharing based on game theory. Law et al. [20] studied the system performance degradation due to the competition of secondary users in distributed spectrum access game. A common assumption of the above results is that each user knows the complete network information to act with the best response strategy. This is, however, often expensive or infeasible to achieve due to significant signaling overhead and the competitors' unwillingness to share information.

To mitigate the strong information requirement for distributed spectrum access, some research results investigate the learning approach for distributed spectrum access such that secondary users adapt the spectrum access decisions locally. Han et al. [21] and Maskery et al. [22] used no-regret learning to solve this problem, assuming that the users' channel selections are common information. The learning converges to a correlated equilibrium [23], wherein the common observed history serves as a signal to coordinate all users' channel selections. When users' channel selections are not observable, authors in [24–26] designed a multi-agent multi-armed bandit learning algorithm to minimize the expected performance loss of distributed spectrum access. Li [27] applied reinforcement learning to analyze Aloha-type spectrum access. Such learning mechanisms relax the strong information requirement by relying on each individual secondary user's local adaption and experience. In a sharp contrast, the proposed social cognitive radio network mechanisms in this book overcome the challenge of limited network information through cooperative social interactions among secondary users.

Only a few efforts have been made to investigate the social intelligence for distributed spectrum sharing. Xing and Chandramouli [28] proposed to use anthropological models in human society to enhance the performance of cognitive radio networks. Li et al. [29] applied the social network approach to analyze the social behavior in cognitive radio networks. Chen et al. [30] proposed a social group utility maximization framework for database-assisted spectrum access such that each user is socially aware and cares about its social friends. In this book, we develop socially inspired distributed spectrum sharing schemes by leveraging three important social mechanisms (i.e., recommendation, imitation, and evoltuion) in human and animal social interactions.

References

1. T. Cisco, Cisco visual networking index: global mobile data traffic forecast update, 2012–2017, in *Cisco Public Information* (2013)
2. P. Goldstein, Credit suisse report: US wireless networks running at 80 % of total capacity, www.FierceWireless.com, July (2011)
3. P. Wendell, M.J. Freedman, Going viral: flash crowds in an open CDN, in *ACM SIGCOMM Conference on Internet Measurement Conference* (2011)
4. F.S.P.T. Force, FCC report of the spectrum efficiency working group, November, 2009
5. M.A. McHenry, D. McCloskey, D. Roberson, J.T. MacDonald, Spectrum occupancy measurements. Technical Report, Shared Spectrum Company, 2005
6. I. Akyildiz, W. Lee, M. Vuran, S. Mohanty, Next generation/dynamic spectrum access/cognitive radio wireless networks: a survey. Comput. Netw. **50**(13), 2127–2159 (2006)
7. D. Sumpter, *Collective Animal Behavior* (Princeton University Press, Princeton, 2010)
8. J. Kennedy, R. Eberhart, Particle swarm optimization, in *IEEE International Conference on Neural Networks*, vol. 4, pp. 1942–1948 (1995)
9. D. Pham, A. Ghanbarzadeh, E. Koc, S. Otri, S. Rahim, M. Zaidi, The bees algorithm—a novel tool for complex optimisation problems, in *IPROMS Conference*, pp. 454–461 (2006)
10. C. Zhang, P. Li, Y. Fang, P. Khargonekar, Decentralized routing in nonhomogeneous poisson networks, in *The International Conference on Distributed Computing Systems (ICDCS)* (2008)
11. M. Brust, C. Ribeiro, D. Turgut, S. Rothkugel, LSWTC: a local small-world topology control algorithm for backbone-assisted mobile ad hoc networks, in *IEEE Conference on Local Computer Networks (LCN)* (2010)

12. N. Nie, C. Comaniciu, Adaptive channel allocation spectrum etiquette for cognitive radio networks, in *First IEEE International Symposium on New Frontiers in Dynamic Spectrum Access Networks* (2005)
13. D. Niyato, E. Hossain, Competitive spectrum sharing in cognitive radio networks: a dynamic game approach. IEEE Trans. Wirel. Commun. **7**(7), 2651–2660 (2008)
14. M. Felegyhazi, M. Cagalj, J.-P. Hubaux, Efficient MAC in cognitive radio systems: a game-theoretic approach. IEEE Trans. Wirel. Commun. **8**(4), 1984–1995 (2009)
15. L. Yang, H. Kim, J. Zhang, M. Chiang, C.W. Tan, Pricing-based spectrum access control in cognitive radio networks with random access, in *IEEE INFOCOM* (2011)
16. D. Li, Y. Xu, J. Liu, X. Wang, Z. Han, A market game for dynamic multi-band sharing in cognitive radio networks, in *IEEE International Conference on Communications (ICC)* (2010)
17. X. Chen, J. Huang, Spatial spectrum access game: Nash equilibria and distributed learning, in *Thirteenth ACM International Symposium on Mobile Ad Hoc Networking and Computing* (2012)
18. X. Chen, J. Huang, Distributed spectrum access with spatial reuse. IEEE J. Sel. Areas Commun. **31**(3), 593–603 (2013)
19. R. Southwell, X. Chen, J. Huang, Quality of service games for spectrum sharing. IEEE J. Sel. Areas Commun. **32**(3), 589–600 (2014)
20. L.M. Law, J. Huang, M. Liu, S.Y. Li et al., Price of anarchy for cognitive MAC games, in *IEEE Global Telecommunications Conference (GLOBECOM)* (2009)
21. Z. Han, C. Pandana, K.R. Liu, Distributive opportunistic spectrum access for cognitive radio using correlated equilibrium and no-regret learning, in *IEEE Wireless Communications and Networking Conference* (2007)
22. M. Maskery, V. Krishnamurthy, Q. Zhao, Decentralized dynamic spectrum access for cognitive radios: cooperative design of a non-cooperative game. IEEE Trans. Commun. **57**(2), 459–469 (2009)
23. R.J. Aumann, Correlated equilibrium as an expression of Bayesian rationality. Econometrica **55**, 1–18 (1987)
24. A. Anandkumar, N. Michael, A. Tang, Opportunistic spectrum access with multiple users: learning under competition, in *IEEE INFOCOM* (2010)
25. L. Lai, H. Jiang, H.V. Poor, Medium access in cognitive radio networks: a competitive multi-armed bandit framework, in *42nd Asilomar Conference on Signals, Systems and Computers* (2008)
26. K. Liu, Q. Zhao, Decentralized multi-armed bandit with multiple distributed players, in *Information Theory and Applications Workshop (ITA)* (2010)
27. H. Li, Multiagent Q-learning for aloha-like spectrum access in cognitive radio systems. EURASIP J. Wirel. Commun. Netw. **2010**(1), 176–216 (2010)
28. Y. Xing, R. Chandramouli, Human behavior inspired cognitive radio network design. IEEE Commun. Mag. **46**(12), 122–127 (2008)
29. H. Li, C.-F. Chen, L. Lai, Propagation of spectrum preference in cognitive radio networks: a social network approach, in *IEEE International Conference on Communications (ICC)* (2011)
30. X. Chen, X. Gong, L. Yang, J. Zhang, A social group utility maximization framework with applications in database assisted spectrum access, in *IEEE INFOCOM* (2014)

Chapter 2
Adaptive Channel Recommendation Mechanism

2.1 Introduction

Designing an efficient spectrum access mechanism for cognitive radio networks is challenging for several reasons: (1) *time-variation*: spectrum opportunities available for secondary users are often time-varying due to primary users' stochastic activities [1]; and (2) *limited observations*: each secondary user often has a limited view of the spectrum opportunities due to the limited spectrum sensing capability [2]. Several characteristics of the wireless channels, on the other hand, turn out to be useful for designing efficient spectrum access mechanisms: (1) *temporal correlations*: spectrum availabilities are correlated in time, and thus observations in the past can be useful in the near future [3]; and (2) *spatial correlation*: secondary users close to one another may experience similar spectrum availabilities [4]. In this chapter, we shall explore the time and space correlations and propose a recommendation-based cooperative spectrum access algorithm, which achieves good communication performances for the secondary users.

Our algorithm design is directly inspired by the recommendation system in the electronic commerce industry. For example, existing owners of various products can provide recommendations (reviews) on Amazon.com, so that other potential customers can pick the products that best suit their needs. Motivated by this, Li [5] proposed a static channel recommendation scheme that encourages secondary users to recommend the channels they have successfully accessed to nearby secondary users. Since each secondary user originally only has a limited view of spectrum availability, such information exchange enables secondary users to take advantages of the correlations in time and space, make more informed decisions, and achieve a high total transmission rate. Similarly as the Geo-location database approach required by FCC for white-space spectrum access [6], we can view the channel recommendation approach as a real-time distributed database generated by the secondary users. This is desirable, for example, when the PU activities change fast (e.g., cellular systems) and a centralized database is difficult to capture the real-time status of all primary users.

© The Author(s) 2015
X. Chen and J. Huang, *Social Cognitive Radio Networks*,
SpringerBriefs in Electrical and Computer Engineering,
DOI 10.1007/978-3-319-15215-8_2

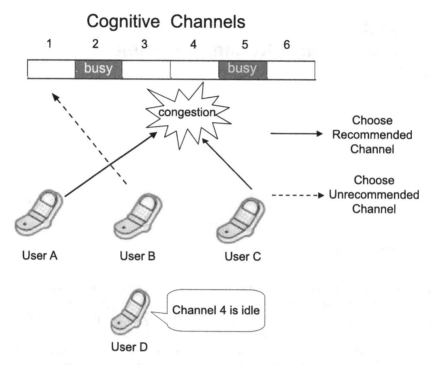

Fig. 2.1 Illustration of the channel recommendation scheme. User D recommends channel 4 to other users. As a result, both user A and user C access the same channel 4, and thus lead to congestion and a reduced rate for both users

The static recommendation scheme in [5], however, ignores two important characteristics of cognitive radios. The first one is the *time variability* we mentioned before. The second one is the *congestion effect*. As depicted in Fig. 2.1, too many users accessing the same channel leads to congestion and a reduced rate for everyone.

To address the shortcomings of the static recommendation scheme, in this chapter we propose an adaptive channel recommendation scheme, which adaptively changes the spectrum access probabilities based on users' latest channel recommendations. We formulate and analyze the system as a Markov decision process (MDP), and propose a numerical algorithm that always converges to the optimal spectrum access policy.

The main results and contributions of this chapter include:

- *Markov decision process formulation*: we formulate and analyze the optimal recommendation-based spectrum access as an average reward MDP.
- *Existence and structure of the optimal policy*: we show that there always exists a stationary optimal spectrum access policy, which requires only the channel recommendation information of the most recent time slot. We also explicitly characterize the structure of the optimal stationary policy with channel homogeneity in

two asymptotic cases (either the number of channels or the number of users goes to infinity).

- *Novel algorithm for finding the optimal policy*: we propose an algorithm based on the recently developed Model Reference Adaptive Search method [7] to find the optimal stationary spectrum access policy. The algorithm has a low complexity even when dealing with a continuous action space of the MDP. We also show that it always converges to the optimal stationary policy. We further propose an efficient heuristic scheme for the heterogeneous channel recommendation, which can significantly reduce the computational time while has small performance loss.
- *Superior performance*: we show that the proposed algorithm achieves up to 18 and 100 % performance improvement than the static channel recommendation scheme in homogeneous and heterogeneous channel environments, respectively, and is also robust to channel dynamics.

The rest of the chapter is organized as follows. We introduce the system model in Sect. 2.2. We then review the static channel recommendation scheme and discuss the motivation for designing an adaptive channel recommendation scheme in Sect. 2.3. The Markov decision process formulation and the structure results of the optimal policy are presented in Sect. 2.4, followed by the Model Reference Adaptive Search based algorithm in Sect. 2.5. We then develop a heuristic scheme for heterogeneous channel recommendation in Sect. 2.6. We illustrate the performance of the algorithms through numerical results in Sect. 2.8 and conclude in Sect. 2.9. *Due to space limitations, the details for several proofs are provided in* [8].

2.2 System Model

We consider a cognitive radio network with M parallel and stochastically heterogeneous primary channels. N homogeneous secondary users try to access these channels using a slotted transmission structure (see Fig. 2.2). The secondary users can exchange information by broadcasting messages over a common control channel.[1] We assume that the secondary users are located close-by, thus they experience similar spectrum availabilities and can hear one another's broadcasting messages. To protect the primary transmissions, secondary users need to sense the channel states before their data transmission.

The system model is described as follows:

- *Channel state*: For each primary channel m, the channel state at time slot t is

$$S_m(t) = \begin{cases} 0, & \text{if channel } m \text{ is occupied by primary transmissions,} \\ 1, & \text{if channel } m \text{ is idle.} \end{cases}$$

---

[1] Please refer to [9] for the details on how to set up and maintain a reliable common control channel in cognitive radio networks.

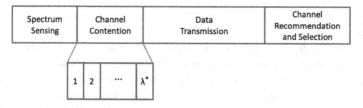

Fig. 2.2 Structure of each spectrum access time slot

Fig. 2.3 Two states
Markovian channel model

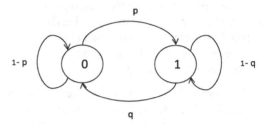

- *Channel state transition*: The states of different channels change according to independent Markovian processes (see Fig. 2.3). We denote the channel state probability vector of channel m at time t as $\boldsymbol{p}_m(t) \triangleq (Pr\{S_m(t) = 0\}, Pr\{S_m(t) = 1\})$, which follows a two-state Markov chain as $\boldsymbol{p}_m(t) = \boldsymbol{p}_m(t-1)\Gamma_m, \forall t \geq 1$, with the transition matrix

$$\Gamma_m = \begin{bmatrix} 1 - p_m & p_m \\ q_m & 1 - q_m \end{bmatrix}.$$

Note that when $p_m = 0$ or $q_m = 0$, the channel state stays unchanged. In the rest of the chapter, we will look at the more interesting and challenging cases where $0 < p_m \leq 1$ and $0 < q_m \leq 1$. The stationary distribution of the Markov chain is given as

$$\lim_{t \to \infty} Pr\{S_m(t) = 0\} = \frac{q_m}{p_m + q_m}, \tag{2.1}$$

$$\lim_{t \to \infty} Pr\{S_m(t) = 1\} = \frac{p_m}{p_m + q_m}. \tag{2.2}$$

- *Heterogeneous channel throughput*: When a secondary user transmits successfully on an idle channel m, it achieves a data rate of B_m. Different channels can support different data rates.
- *Channel contention*: To resolve the transmission collision when multiple secondary users access the same channel, a backoff mechanism is used (see Fig. 2.2 for illustration). The contention stage of a time slot is divided into λ^* mini-slots, and each user n executes the following two steps:

1. Count down according to a randomly and uniformly chosen integral backoff time (number of mini-slots) λ_n between 1 and λ^*.
2. Once the timer expires, monitor the channel and transmit RTS/CTS messages to grab the channel if the channel is clear (i.e., no ongoing transmission). Note that if multiple users choose the same backoff mini-slot, a collision will occur with RTS/CTS transmissions and no users can grab the channel. Once successfully grabing the channel, the user starts to transmit its data packet.

Suppose that k_m users choose channel m to access. Then the probability that user n (out of the k_m users) successfully grabs the channel m is

$$Pr_n = Pr\{\lambda_n < \min_{i \neq n}\{\lambda_i\}\}$$

$$= \sum_{\lambda=1}^{\lambda^*} Pr\{\lambda_n = \lambda\}Pr\{\lambda < \min_{i \neq n}\{\lambda_i\}|\lambda_n = \lambda\}$$

$$= \sum_{\lambda=1}^{\lambda^*} \frac{1}{\lambda^*}\left(\frac{\lambda^* - \lambda}{\lambda^*}\right)^{k_m-1}. \tag{2.3}$$

For the ease of exposition, we will focus on the asymptotic case where λ^* goes to ∞. This is a good approximation when the number of mini-slots λ^* for backoff is much larger than the number of users N and collisions rarely occur. It simplifies the analysis as

$$\lim_{\lambda^* \to \infty} \sum_{\lambda=1}^{\lambda^*} \frac{1}{\lambda^*}\left(\frac{\lambda^* - \lambda}{\lambda^*}\right)^{k_m-1} = \lim_{\frac{1}{\lambda^*} \to 0} \sum_{\lambda=0}^{\lambda^*-1} \left(\frac{\lambda}{\lambda^*}\right)^{k_m-1}\frac{1}{\lambda^*}$$

$$= \int_0^1 z^{k_m-1}dz = \frac{1}{k_m}, \tag{2.4}$$

and thus the expected throughput of user n is

$$u_n(t) = \frac{B_m S_m(t)}{k_m}. \tag{2.5}$$

In Sect. 2.7, we also generalize the results to the case that $\lambda^* < \infty$.

2.3 Introduction to Channel Recommendation

In this section, we first give a review of the static channel recommendation scheme in [5] and then discuss the motivation for adaptive channel recommendation.

2.3.1 Review of Static Channel Recommendation

The key idea of the static channel recommendation scheme is that secondary users inform each other about the available channels they have just accessed. More specifically, each secondary user executes the following four stages synchronously during each time slot (See Fig. 2.2):

- *Spectrum sensing*: sense one of the channels based on channel selection result made at the end of the previous time slot.
- *Channel contention*: if the channel sensing result is idle, compete for the channel with the backoff mechanism described in Sect. 2.2.
- *Data transmission*: transmit data packets if the user successfully grabs the channel.
- *Channel recommendation and selection*:

  - *Announce recommendation*: if the user has successfully accessed an idle channel, broadcast this channel ID to all other secondary users.
  - *Collect recommendation*: collect recommendations from other secondary users and store them in a buffer. Typically, the correlation of channel availabilities between two slots diminishes as the time difference increases. Therefore, each secondary user will only keep the recommendations received from the most recent W slots and discard the out-of-date information. The user's own successful transmission history within W recent time slots is also stored in the buffer. W is a system design parameter and will be further discussed later.
  - *Select channel*: choose a channel to sense at the next time slot by putting more weights on the recommended channels according to a *static branching probability* P_{rec}. Suppose that the user has $0 < R < M$ different channel recommendations in the buffer, then the probability of accessing a channel m is

$$P_m = \begin{cases} \frac{P_{rec}}{R}, & \text{if channel } m \text{ is recommended,} \\ \frac{1 - P_{rec}}{M - R}, & \text{otherwise.} \end{cases} \tag{2.6}$$

A larger value of P_{rec} means that putting more weight on the recommended channels. When $R = 0$ (no channel is recommended) or M (all channels are recommended), the random access is used and the probability of selecting channel m is $P_m = \frac{1}{M}$.

To illustrate the channel selection process, let us take the network in Fig. 2.1 as an example. Suppose that the branching probability $P_{rec} = 0.4$. Since only $R = 1$ recommendation is available (i.e., channel 4), the probabilities of choosing the recommended channel 4 and any unrecommended channel are $\frac{0.4}{1} = 0.4$ and $\frac{1-0.4}{6-1} = 0.12$, respectively.

Numerical studies in [5] showed that the static channel recommendation scheme achieves a higher performance over the traditional random channel access scheme without information exchange. However, the fixed value of P_{rec} limits the performance of the static scheme, as explained next.

2.3.2 Motivations for Adaptive Channel Recommendation

The static channel recommendation mechanism is simple to implement due to a fixed value of P_{rec}. However, it may lead to significant congestions when the number of recommended channels is small. In the extreme case when only $R = 1$ channel is recommended, calculation (2.6) suggests that every user will access that channel with a probability P_{rec}. When the number of users N is large, the expected number of users accessing this channel $N P_{rec}$ will be high. Thus heavy congestion happens and each secondary user will get a low expected throughput.

A better way is to adaptively change the value of P_{rec} based on the number of recommended channels. This is the key idea of our proposed algorithm. To illustrate the advantage of adaptive algorithms, let us first consider a simple heuristic adaptive algorithm in a homogeneous channel environment, i.e., for each channel m, its data rate $B_m = B$ and channel state changing probabilities $p_m = p, q_m = q$. In this algorithm, we choose the branching probability such that the expected number of secondary users choosing a single recommended channel is one. To achieve this, we need to set P_{rec} as in Lemma 2.1.

Lemma 2.1 *If we choose the branching probability $P_{rec} = \frac{R}{N}$, then the expected number of secondary users choosing any one of the R recommended channels is one.*

Without going through detailed analysis, it is straightforward to show the benefit for such adaptive approach through simple numerical examples. Let us consider a network with $M = 10$ channels and $N = 5$ secondary users. For each channel m, the initial channel state probability vector is $\boldsymbol{p}_m(0) = (0, 1)$ and the transition matrix is

$$\Gamma_m = \begin{bmatrix} 1 - 0.01\varepsilon & 0.01\varepsilon \\ 0.01\varepsilon & 1 - 0.01\varepsilon \end{bmatrix},$$

where ε is called the dynamic factor. A larger value of ε implies that the channels are more dynamic over time. We are interested in time average system throughput $U = \frac{\sum_{t=1}^{T} \sum_{n=1}^{N} u_n(t)}{T}$, where $u_n(t)$ is the throughput of user n at time slot t. In the simulation, we set the total number of time slots $T = 2,000$.

We implement the following three channel access schemes:

- Random access scheme: each secondary user selects a channel randomly.
- Static channel recommendation scheme as in [5] with the *optimal* constant branching probability $P_{rec} = 0.7$.
- Heuristic adaptive channel recommendation scheme with the variable branching probability $P_{rec} = \frac{R}{N}$.

Figure 2.4 shows that the heuristic adaptive channel recommendation scheme outperforms the static channel recommendation scheme, which in turn outperforms the random access scheme. Moreover, the heuristic adaptive scheme is more robust to the dynamic channel environment, as it decreases slower than the static scheme when ε increases.

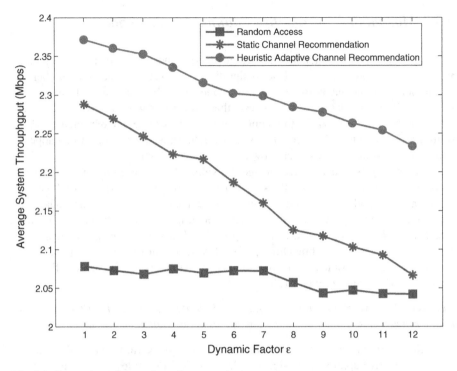

Fig. 2.4 Comparison of three channel access schemes

We can imagine that an optimal adaptive scheme (by setting the right $P_{rec}(t)$ over time) can further increase the network performance. However, computing the optimal branching probability in closed-form is very difficult. In the rest of the chapter, we will focus on characterizing the structures of the optimal spectrum access strategy and designing an efficient algorithm to achieve the optimum.

2.4 Adaptive Channel Recommendation with Channel Homogeneity

We first study the optimal channel recommendation in the homogeneous channel environment, i.e., each channel m has the same data rate $B_m = B$ and identical channel state changing probabilities $p_m = p, q_m = q$. The generalization to the heterogeneous channel setting will be discussed in Sect. 2.6. To find the optimal adaptive spectrum access strategy, we formulate the system as a Markov Decision Process (MDP). For the sake of simplicity, we assume that the recommendation buffer size $W = 1$, i.e., users only consider the recommendations received in the last time slot. Our method also applies to the case when $W > 1$ by using a high-order MDP formulation, although the analysis is more involved.

2.4.1 MDP Formulation for Adaptive Channel Recommendation

We model the system as a MDP as follows:

- *System state*: $R \in \mathscr{R} \triangleq \{0, 1, \ldots, \min\{M, N\}\}$ denotes the number of recommended channels at the end of time slot t. Since all channels are statistically homogenous, then there is no need to keep track of the recommended channel IDs.
- *Action*: $P_{rec} \in \mathscr{P} \triangleq (0, 1)$ denotes the branching probability of choosing the set of recommended channels.
- *Transition probability*: The probability that action P_{rec} in system state R in time slot t will lead to system state R' in the next time slot is $P_{R,R'}^{P_{rec}} = Pr\{R(t+1) = R'|R(t) = R, P_{rec}(t) = P_{rec}\}$. We can compute this probability as in (2.7), with detailed derivations given in [8].

$$
P_{R,R'}^{P_{rec}} = \sum_{m_r+m_u=R'} \sum_{R \geq \bar{m}_r \geq m_r, M-R \geq \bar{m}_u \geq m_u} \sum_{n_r+n_u=N, n_r \geq \bar{m}_r, n_u \geq \bar{m}_u} \binom{N}{n_r}
$$
$$
\times P_{rec}^{n_r}(1 - P_{rec})^{n_u} \binom{\bar{m}_r}{m_r} (1-q)^{m_r} q^{\bar{m}_r - m_r} \frac{R!}{(R-\bar{m}_r)!} \binom{n_r-1}{\bar{m}_r-1} R^{-n_r}
$$
$$
\times \binom{\bar{m}_u}{m_u} \left(\frac{p}{p+q}\right)^{m_u} \left(\frac{q}{p+q}\right)^{\bar{m}_u - m_u} \frac{(M-R)!}{(M-R-\bar{m}_u)!} \binom{n_u-1}{\bar{m}_u-1} (M-R)^{-n_u}.
$$

$$(2.7)$$

- *Reward*: $U(R, P_{rec})$ is the expected system throughput in next time slot when the action P_{rec} is taken in current system state R, i.e., $U(R, P_{rec}) = \sum_{R' \in \mathscr{R}} P_{R,R'}^{P_{rec}} U_{R'}$, where $U_{R'}$ is the system throughput in state R'. If R' idle channels are utilized by the secondary users in a time slot, then these R' channels will be recommended at the end of the time slot. Thus, we have $U_{R'} = R'B$. Recall that B is the data rate that a single user can obtain on an idle channel.
- *Stationary policy*: $\pi \in \Omega \triangleq \mathscr{P}^{|\mathscr{R}|}$ maps from each state R to an action P_{rec}, i.e., $\pi(R)$ is the action P_{rec} taken when the system is in state R. The mapping is stationary and does not depend on time t.

Given a stationary policy π and the initial state $R_0 \in \mathscr{R}$, we define the network's value function as the time average system throughput, i.e.,

$$
\Phi_\pi(R_0) = \lim_{T \to \infty} \frac{1}{T} E_\pi \left[\sum_{t=0}^{T-1} U(R(t), \pi(R(t))) \right].
$$

We want to find an optimal stationary policy π^* that maximizes the value function $\Phi_\pi(R_0)$ for any initial state R_0, i.e., $\pi^* = \arg\max_\pi \Phi_\pi(R_0), \forall R_0 \in \mathscr{R}$. Notice that this is a system wide optimization, although the optimal solution can be implemented in a distributed fashion. For example, each user can calculate the optimal spectrum access policy off-line, and determine the real-time optimal channel access probability

P_{rec} locally by observing the number of recommended channels R after entering the network.

2.4.2 Existence of Optimal Stationary Policy

MDP formulation above is an average reward based MDP. We show in Theorem 2.1 that an optimal stationary policy that is independent of initial system state always exists in our MDP formulation.

Theorem 2.1 *There exists an optimal stationary policy for the adaptive channel recommendation MDP.*

Furthermore, the optimal stationary policy π^* is independent of the initial state R_0 due to the irreducibility of the adaptive channel recommendation MDP, i.e., $\Phi_{\pi^*}(R_0) = \Phi_{\pi^*}, \forall R_0 \in \mathcal{R}$, where Φ_{π^*} is the maximum time average system throughput. In the rest of the chapter, we will just use "optimal policy" to refer "optimal stationary policy that is independent of the initial system state".

2.4.3 Structure of Optimal Stationary Policy

Next we characterize the structure of the optimal policy without using the closed-form expressions of the policy (which is generally hard to achieve). The key idea is to treat the average reward based MDPs as the limit of a sequence of discounted reward MDPs with discounted factors going to one. Under the irreducibility condition, the average reward based MDP thus inherits the structure property from the corresponding discounted reward MDP [10]. We can write down the Bellman equations of the discounted version of our MDP problem as:

$$V_t(R) = \max_{P_{rec} \in \mathcal{P}} \sum_{R' \in \mathcal{R}} P_{R,R'}^{P_{rec}} [U_{R'} + \beta V_{t+1}(R')], \quad \forall R \in \mathcal{R}, \qquad (2.8)$$

where $V_t(R)$ is the discounted maximum expected system throughput starting from time slot t when the system in state R, and $0 < \beta < 1$ is the discounted factor.

Due to the combinatorial complexity of the transition probability $P_{R,R'}^{P_{rec}}$ in (2.7), it is difficult to obtain the structure results for the general case. We further limit our attention to the following two asymptotic cases.

2.4.3.1 Case One: The Number of Channels M Goes to Infinity While the Number of Users N Stays Finite

In this case, the number of channels is much larger than the number of secondary users, and thus heavy congestion rarely happens on any channel. Thus it is safe to emphasizing on accessing the recommended channels. Before proving the main result of Case One in Theorem 2.2, let us first characterize the property of discounted maximum expected system payoff $V_t(R)$.

Proposition 2.1 *When $M = \infty$ and $N < \infty$, the value function $V_t(R)$ for the discounted adaptive channel recommendation MDP is nondecreasing in R.*

Based on the monotone property of the value function $V_t(R)$, we prove the following main result.

Theorem 2.2 *When $M = \infty$ and $N < \infty$, for the adaptive channel recommendation MDP, the optimal stationary policy π^* is monotone, that is, $\pi^*(R)$ is nondecreasing on $R \in \mathscr{R}$.*

2.4.3.2 Case Two: The Number of Users N Goes to Infinity While the Number of Channels M Stays Finite

In this case, the number of secondary users is much larger than the number of channels, and thus congestion becomes a major concern. However, since there are infinitely many secondary users, all the idle channels at each time slot can be utilized as long as users have positive probabilities to access all channels. From the system's point of view, the cognitive radio network operates in the saturation state. Formally, we show that

Theorem 2.3 *When $N = \infty$ and $M < \infty$, for the adaptive channel channel recommendation MDP, any stationary policy π satisfying $0 < \pi(R) < 1, \forall R \in \mathscr{R}$ is optimal.*

2.5 Model Reference Adaptive Search for Optimal Spectrum Access Policy

Next we will design an algorithm that can converge to the optimal policy under general system parameters (not limiting to the two asymptotic cases). Since the action space of the adaptive channel recommendation MDP is continuous (i.e., choosing a probability P_{rec} in $(0, 1)$), the traditional method of discretizing the action space followed by the policy, value iteration, or Q-learning cannot guarantee to converge to the optimal policy. To overcome this difficulty, we propose a new algorithm developed from the Model Reference Adaptive Search method, which was recently developed in the Operations Research community [7]. We will show that the proposed algorithm is easy to implement and is provably convergent to the optimal policy.

2.5.1 Model Reference Adaptive Search Method

We first introduce the basic idea of the Model Reference Adaptive Search (MRAS) method. Later on, we will show how the method can be used to obtain optimal spectrum access policy for our problem.

The MRAS method is a new randomized method for global optimization [7]. The key idea is to randomize the original optimization problem over the feasible region according to a specified probabilistic model. The method then generates candidate solutions and updates the probabilistic model on the basis of elite solutions and a reference model, so that to guide the future search toward better solutions.

Formally, let $J(x)$ be the objective function to maximize. The MRAS method is an iterative algorithm, and it includes three phases in each iteration k:

- *Random solution generation*: generate a set of random solutions $\{x\}$ in the feasible set χ according to a parameterized probabilistic model $f(x, v_k)$, which is a probability density function (pdf) with parameter v_k. The number of solutions to generate is a fixed system parameter.
- *Reference distribution construction*: select elite solutions among the randomly generated set, such that the chosen ones satisfy $J(x) \geq \gamma$. Construct a reference probability distribution as

$$
g_k(x) = \begin{cases} \dfrac{I_{\{J(x)\geq\gamma\}}}{E_{f(x,v_0)}[\frac{I_{\{J(x)\geq\gamma\}}}{f(x,v_0)}]} & k = 1, \\[2em] \dfrac{e^{J(x)}I_{\{J(x)\geq\gamma\}}g_{k-1}(x)}{E_{g_{k-1}}[e^{J(x)}I_{\{J(x)\geq\gamma\}}]} & k \geq 2, \end{cases}
\tag{2.9}
$$

where $I_{\{\varpi\}}$ is an indicator function, which equals 1 if the event ϖ is true and zero otherwise. Parameter v_0 is the initial parameter for the probabilistic model (used during the first iteration, i.e., $k = 1$), and $g_{k-1}(x)$ is the reference distribution in the previous iteration (used when $k \geq 2$).

- *Probabilistic model update*: update the parameter v of the probabilistic model $f(x, v)$ by minimizing the Kullback-Leibler divergence between $g_k(x)$ and $f(x, v)$, i.e.,

$$
v_{k+1} = \arg\min_v E_{g_k}\left[\ln \frac{g_k(x)}{f(x, v)}\right].
\tag{2.10}
$$

By constructing the reference distribution according to (2.9), the expected performance of random elite solutions can be improved under the new reference distribution, i.e.,

$$
\begin{aligned}
E_{g_k}[e^{J(x)}I_{\{J(x)\geq\gamma\}}] &= \frac{\int_{x\in\chi} e^{2J(x)}I_{\{J(x)\geq\gamma\}}g_{k-1}(x)dx}{E_{g_{k-1}}[e^{J(x)}I_{\{J(x)\geq\gamma\}}]} \\
&= \frac{E_{g_{k-1}}[e^{2J(x)}I_{\{J(x)\geq\gamma\}}]}{E_{g_{k-1}}[e^{J(x)}I_{\{J(x)\geq\gamma\}}]} \geq E_{g_{k-1}}[e^{J(x)}I_{\{J(x)\geq\gamma\}}].
\end{aligned}
\tag{2.11}
$$

To find a better solution to the optimization problem, it is natural to update the probabilistic model (from which random solution are generated in the first stage) to as close to the new reference probability as possible, as done in the third stage.

2.5.2 Model Reference Adaptive Search for Optimal Spectrum Access Policy

In this section, we design an algorithm based on the MRAS method to find the optimal spectrum access policy. Here we treat the adaptive channel recommendation MDP as a global optimization problem over the policy space. The key challenge is the choice of proper probabilistic model $f(\cdot)$, which is crucial for the convergence of the MRAS algorithm.

2.5.2.1 Random Policy Generation

To apply the MRAS method, we first need to set up a random policy generation mechanism. Since the action space of the channel recommendation MDP is continuous, we use the Gaussian distributions. Specifically, we generate sample actions $\pi(R)$ from a Gaussian distribution for each system state $R \in \mathscr{R}$ independently, i.e. $\pi(R) \sim \mathscr{N}(\mu_R, \sigma_R^2)$.[2] In this case, a candidate policy π can be generated from the joint distribution of $|\mathscr{R}|$ independent Gaussian distributions, i.e.,

$$(\pi(0), \ldots, \pi(\min\{M, N\})) \sim \mathscr{N}(\mu_0, \sigma_0^2) \times \cdots \times \mathscr{N}(\mu_{\min\{M,N\}}, \sigma_{\min\{M,N\}}^2).$$

As shown later, Gaussian distribution has nice analytical and convergent properties for the MRAS method.

For the sake of brevity, we denote $f(\pi(R), \mu_R, \sigma_R)$ as the pdf of the Gaussian distribution $\mathscr{N}(\mu_R, \sigma_R^2)$, and $f(\pi, \mu, \sigma)$ as random policy generation mechanism with parameters $\mu \triangleq (\mu_0, \ldots, \mu_{\min\{M,N\}})$ and $\sigma \triangleq (\sigma_0, \ldots, \sigma_{\min\{M,N\}})$, i.e.,

$$f(\pi, \mu, \sigma) = \prod_{R=0}^{\min\{M,N\}} f(\pi(R), \mu_R, \sigma_R)$$

$$= \prod_{R=0}^{\min\{M,N\}} \frac{1}{\sqrt{2\varphi\sigma_R^2}} e^{-\frac{(\pi(R)-\mu_R)^2}{2\sigma_R^2}},$$

where φ is the circumference-to-diameter ratio.

---

[2] Note that the Gaussian distribution has a support over $(-\infty, +\infty)$, which is larger than the feasible region of $\pi(R)$. This issue will be handled in Sect. 2.5.2.2.

2.5.2.2 System Throughput Evaluation

Given a candidate policy π randomly generated based on $f(\pi, \boldsymbol{\mu}, \boldsymbol{\sigma})$, we need to evaluate the expected system throughput Φ_π. From (2.7), we obtain the transition probabilities $P_{R,R'}^{\pi(R)}$ for any system state $R, R' \in \mathscr{R}$. Since a policy π leads to a finitely irreducible Markov chain, we can obtain its stationary distribution. Let us denote the transition matrix of the Markov chain as $Q \triangleq [P_{R,R'}^{\pi(R)}]_{|\mathscr{R}| \times |\mathscr{R}|}$ and the stationary distribution as $\boldsymbol{p} = (Pr(0), \ldots, Pr(\min\{M, N\}))$. Obviously, the stationary distribution can be obtained by solving the equation $\boldsymbol{p}Q = \boldsymbol{p}$. We then calculate the expected system throughput Φ_π by $\Phi_\pi = \sum_{R \in \mathscr{R}} Pr(R)U_R$.

Note that in the discussion above, we assume that $\pi \in \Omega$ implicitly, where Ω is the feasible policy space. Since Gaussian distribution has a support over $(-\infty, +\infty)$, we thus extend the definition of expected system throughput Φ_π over $(-\infty, +\infty)^{|\mathscr{R}|}$ as

$$
\Phi_\pi = \begin{cases} \sum_{R \in \mathscr{R}} Pr(R)U_R & \pi \in \Omega, \\ -\infty & \text{Otherwise.} \end{cases}
$$

In this case, whenever any generated policy π is not feasible, we have $\Phi_\pi = -\infty$. As a result, such policy π will not be selected as an elite sample (discussed next) and will not be used for probability updating. Hence the search of MRAS algorithm will not bias towards any unfeasible policy space.

2.5.2.3 Reference Distribution Construction

To construct the reference distribution, we first need to select the elite policies. Suppose L candidate policies, $\pi_1, \pi_2, \ldots, \pi_L$, are generated at each iteration. We order them based on an increasing order of the expected system throughputs Φ_π, i.e., $\Phi_{\hat{\pi}_1} \leq \Phi_{\hat{\pi}_2} \leq \ldots \leq \Phi_{\hat{\pi}_L}$, and set the elite threshold as $\gamma = \Phi_{\hat{\pi}_{\lceil (1-\rho)L \rceil}}$, where $0 < \rho < 1$ is the elite ratio. For example, when $L = 100$ and $\rho = 0.4$, then $\gamma = \Phi_{\hat{\pi}_{60}}$ and the last 40 samples in the sequence will be selected as elite samples. Note that as long as L is sufficiently large, we shall have $\gamma < \infty$ and hence only feasible policies π are selected. According to (2.9), we then construct the reference distribution as

$$
g_k(\pi) = \begin{cases} \dfrac{I_{\{\Phi_\pi \geq \gamma\}}}{E_{f(\pi, \mu_0, \sigma_0)}[\frac{I_{\{\Phi_\pi \geq \gamma\}}}{f(\pi, \mu_0, \sigma_0)}]} & k = 1, \\ \dfrac{e^{\Phi_\pi} I_{\{\Phi_\pi \geq \gamma\}} g_{k-1}(\pi)}{E_{g_{k-1}}[e^{\Phi_\pi} I_{\{\Phi_\pi \geq \gamma\}}]} & k \geq 2. \end{cases} \tag{2.12}
$$

2.5.2.4 Policy Generation Update

For the MRAS algorithm, the critical issue is the updating of random policy generation mechanism $f(\pi, \boldsymbol{\mu}, \boldsymbol{\sigma})$, or solving the problem in (2.10). The optimal update rule is described as follow.

Theorem 2.4 *The optimal parameter* (μ, σ) *that minimizes the Kullback-Leibler divergence between the reference distribution* $g_k(\pi)$ *in (2.12) and the new policy generation mechanism* $f(\pi, \mu, \sigma)$ *is*

$$\mu_R = \frac{\int_{\pi \in \Omega} e^{(k-1)\Phi_\pi} I_{\{\Phi_\pi \geq \gamma\}} \pi(R) d\pi}{\int_{\pi \in \Omega} e^{(k-1)\Phi_\pi} I_{\{\Phi_\pi \geq \gamma\}} d\pi}, \quad \forall R \in \mathscr{R}, \tag{2.13}$$

$$\sigma_R^2 = \frac{\int_{\pi \in \Omega} e^{(k-1)\Phi_\pi} I_{\{\Phi_\pi \geq \gamma\}} [\pi(R) - \mu_R]^2 d\pi}{\int_{\pi \in \Omega} e^{(k-1)\Phi_\pi} I_{\{\Phi_\pi \geq \gamma\}} d\pi}, \quad \forall R \in \mathscr{R}. \tag{2.14}$$

2.5.2.5 MARS Algorithm for Optimal Spectrum Access Policy

Based on the MARS algorithm, we generate L candidate polices at each iteration. Then the updates in (2.13) and (2.14) are replaced by the sample average version in (2.15) and (2.16) in Algorithm 1, respectively. As a summary, we describe the MARS-based algorithm for finding the optimal spectrum access policy of adaptive channel recommendation MDP in Algorithm 1.

We then analyze the computational complexity of the MRAS algorithm. For each iteration, the sample generation in Line 4 in Algorithm 1 involves L samples with each generated from $|\mathscr{R}|$ Gaussian distributions. This step has the complexity of $\mathscr{O}(L|\mathscr{R}|)$. The elite sample selection in Line 5 involves the sorting operation, which typically has the complexity of $\mathscr{O}(L \ln L)$. The update in Line 6 involving the summation operation also has the complexity of $\mathscr{O}(L|\mathscr{R}|)$. Suppose that it takes Z iterations for the algorithm to converge. Then the total computational complexity of the MRAS algorithm is $\mathscr{O}(ZL|\mathscr{R}| + ZL \ln L)$.

Algorithm 1 MRAS-based Algorithm For Adaptive Recommendation Based Optimal Spectrum Access

---

1: **initialize** parameters for Gaussian distributions (μ_0, σ_0), the elite ratio ρ, and the stopping criterion ξ. Set initial elite threshold $\gamma_0 = 0$ and iteration index $k = 0$.

2: **repeat**:

3:    **increase** iteration index k by 1.

4:    **generate** L candidate policies $\pi_1, \ldots, \pi_L$ from the random policy generation mechanism $f(\pi, \mu_{k-1}, \sigma_{k-1})$.

5:    **select** elite policies by setting the elite threshold $\gamma_k = \max\{\Phi_{\hat{\pi}_{\lceil(1-\rho)L\rceil}}, \gamma_{k-1}\}$.

6:    **update** the random policy generation mechanism by (for any $\forall R \in \mathscr{R}$)

$$\mu_{R,k} = \frac{\sum_{i=1}^{L} e^{(k-1)\Phi_\pi} I_{\{\Phi_{\pi_i} \geq \gamma_k\}} \pi_i(R)}{\sum_{i=1}^{L} e^{(k-1)\Phi_\pi} I_{\{\Phi_{\pi_i} \geq \gamma_k\}}}, \tag{2.15}$$

$$\sigma_{R,k}^2 = \frac{\sum_{i=1}^{L} e^{(k-1)\Phi_\pi} I_{\{\Phi_{\pi_i} \geq \gamma_k\}} [\pi_i(R) - \mu_R]^2}{\sum_{i=1}^{L} e^{(k-1)\Phi_\pi} I_{\{\Phi_{\pi_i} \geq \gamma_k\}}}. \tag{2.16}$$

7: **until** $\max_{R \in \mathscr{R}} \sigma_{R,k} < \xi$.

2.5.3 Convergence of Model Reference Adaptive Search

In this part, we discuss the convergence property of the MRAS-based optimal spectrum access policy. For ease of exposition, we assume that the adaptive channel recommendation MDP has a unique global optimal policy. Numerical studies in [7] show that the MRAS method also converges for the multiple global optimums case. We shall show that the random policy generation mechanism $f(\pi, \mu_k, \sigma_k)$ will eventually generate the optimal policy.

Theorem 2.5 *For the MRAS algorithm, the limiting point of the policy sequence $\{\pi_k\}$ generated by the sequence of random policy generation mechanism $\{f(\pi, \mu_k, \sigma_k)\}$ converges point-wisely to the optimal spectrum access policy π^* for the adaptive channel recommendation MDP, i.e.,*

$$\lim_{k\to\infty} E_{f(\pi,\mu_k,\sigma_k)}[\pi(R)] = \pi^*(R), \quad \forall R \in \mathscr{R}, \tag{2.17}$$

$$\lim_{k\to\infty} Var_{f(\pi,\mu_k,\sigma_k)}[\pi(R)] = 0, \quad \forall R \in \mathscr{R}. \tag{2.18}$$

From Theorem 2.5, we see that parameter $(\mu_{R,k}, \sigma_{R,k})$ for updating in (2.15) and (2.16) also converges, i.e.,

$$\lim_{k\to\infty} \mu_{R,k} = \pi^*(R), \quad \forall R \in \mathscr{R},$$

$$\lim_{k\to\infty} \sigma_{R,k} = 0, \quad \forall R \in \mathscr{R}.$$

Thus, we can use $\max_{R\in\mathscr{R}} \sigma_{R,k} < \xi$ as the stopping criterion in Algorithm 1.

2.6 Adaptive Channel Recommendation with Channel Heterogeneity

We now generalize the adaptive channel recommendation to the heterogeneous channel setting. Recall that the system state R in the homogeneous channel case only keeps track of how many channels are recommended. In a heterogeneous channel environment, each channel has different a data rate B_m and channel state changing probabilities p_m and q_m. Keeping track of the number of recommend channels is not enough for optimal decision. Intuitively, if a channel with higher data rate B_m is recommended, users should choose this channel with a higher weight. The new system state for the heterogeneous channel case should be defined as a vector $\mathbf{R} \triangleq (I_1, \ldots, I_M)$, where $I_m = 1$ if channel m is recommended and $I_m = 0$ otherwise. The objective of the heterogeneous channel recommendation MDP is then to find the optimal channel access probabilities $\{P_m(\mathbf{R})\}_{m=1}^M$ for each system state $\mathbf{R}$ where $P_m(\mathbf{R})$ is the probability of selecting channel m.

Similarly with the homogeneous channel case, we can apply the MRAS method (by replacing system state R and decision variables P_{rec} in Algorithm 1 with $\mathbf{R}$ and $\{P_m(\mathbf{R})\}_{m=1}^M$, respectively) to obtain the optimal solutions with the new formulation.

However, the number of decision variables $\{P_m(\mathbf{R})\}_{m=1}^M$ in the heterogeneous channel model equals to $M2^M$, which causes exponential blow up in the computational complexity (i.e., $\mathscr{O}\left(ZLM2^M + ZL\ln L\right)$ with the similar analysis as in Sect. 2.5.2.5). We next focus on developing a low complexity efficient heuristic algorithm to solve the MDP.

Recall that in the heuristic algorithm in Lemma 2.1 for the homogeneous channel recommendation, the weight of selecting each recommended channel is $\frac{1}{N}$ and total weights of choosing recommended channels are $R\frac{1}{N}$. Similarly, we can design a low complexity heuristic algorithm for the heterogeneous channel recommendation. More specifically, we set the weight of selecting channel m is P_1^m (P_0^m, respectively) when the channel is recommended (the channel is not recommended, respectively). Given the system is in state $\mathbf{R}$, the probability of choosing channel m is proportional to its weight of its state I_m, i.e.,

$$P_m(\mathbf{R}) = \frac{P_{I_m}^m}{\sum_{m'=1}^M P_{I_{m'}}^m}. \tag{2.22}$$

In this case, the total number of decision variables $P_{I_m}^m$ is reduced to $2M$, which grows linearly in the number of channels M. Let $\boldsymbol{\pi} = \{(P_1^m, P_0^m)\}_{m=1}^M \in (0, 1)^{2M}$ denote the set of corresponding decision variables. Our objective is to find the optimal $\boldsymbol{\pi}$ that maximizes the time average throughput $\Phi_{\boldsymbol{\pi}}$. We can again apply the MRAS method to find the optimal solution, which is given in Algorithm 2. The procedures of derivation is very similar with the MRAS method for the homogeneous channel recommendation; we omit the details due to space limit. With the similar analysis as in Sect. 2.5.2.5, we see that the heuristic algorithm has the computational complexity of $\mathscr{O}(ZLM + ZL\ln L)$.

Algorithm 2 MRAS-based Algorithm For Optimizing Heuristic Heterogeneous Channel Recommendation

---

1: **initialize** parameters for the elite ratio ρ, Gaussian distributions $\mu(0) = \{(\mu_1^m(0), \mu_0^m(0))\}_{m=1}^M$, $\sigma(0) = \{(\sigma_1^m(0), \sigma_0^m(0))\}_{m=1}^M$, and the stopping criterion ξ. Set initial elite threshold $\gamma_0 = 0$ and iteration index $k = 0$.

2: **repeat**:

3:     **increase** iteration index k by 1.

4:     **generate** L candidate policies $\boldsymbol{\pi}_1, \ldots, \boldsymbol{\pi}_L$ from the random policy generation mechanism $f(\boldsymbol{\pi}, \mu(k-1), \sigma(k-1))$.

5:     **select** elite policies by setting the elite threshold $\gamma_k = \max\{\Phi_{\hat{\boldsymbol{\pi}}_{\lceil(1-\rho)L\rceil}}, \gamma_{k-1}\}$.

6:     **update** the random policy generation mechanism by (for any $I_m \in \{0, 1\}, m \in \mathscr{M}$)

$$\mu_{I_m}^m(k) = \frac{\sum_{i=1}^L e^{(k-1)\Phi_{\boldsymbol{\pi}}} I_{\{\Phi_{\boldsymbol{\pi}_i} \geq \gamma_k\}} P_{I_m}^m}{\sum_{i=1}^L e^{(k-1)\Phi_{\boldsymbol{\pi}}} I_{\{\Phi_{\boldsymbol{\pi}_i} \geq \gamma_k\}}}, \tag{2.20}$$

$$\sigma_{I_m}^m(k) = \left(\frac{\sum_{i=1}^L e^{(k-1)\Phi_{\boldsymbol{\pi}}} I_{\{\Phi_{\boldsymbol{\pi}_i} \geq \gamma_k\}} (P_{I_m}^m - \mu_{I_m}^m(k))^2}{\sum_{i=1}^L e^{(k-1)\Phi_{\boldsymbol{\pi}}} I_{\{\Phi_{\boldsymbol{\pi}_i} \geq \gamma_k\}}}\right)^{\frac{1}{2}}. \tag{2.21}$$

7: **until** $\max_{I_m \in \{0,1\}, m \in \mathscr{M}} \sigma_{I_m}^m(k) < \xi$.

Note that the optimal policy π^* for the heuristic heterogeneous channel recommendation is also a feasible policy for the heterogeneous channel recommendation MDP. The performance of the optimal policy for the heterogeneous channel recommendation MDP thus dominates the heuristic heterogeneous channel recommendation. However, numerical results show that the heuristic heterogeneous channel recommendation has a small performance loss comparing to the optimal policy while gaining a significant computation complexity reduction.

2.7 Adaptive Channel Recommendation in General Channel Environment

For the ease of exposition, we consider the Markovian channel model in the analysis above. Such a channel model can be a good approximation of reality if the primary traffic is highly bursty [11]. We now extend the MRAS-based channel recommendation algorithm to a general channel environment including the non-Markovian setting, where it is difficult to obtain the statistical properties apriori.

The key idea is to cast the system throughput optimization problem in the general channel environment as a stochastic optimization problem. Let $\mathbf{S} = (S_1, \ldots, S_M)$ be the states of all channels, which is a random vector generated from a general probability distribution ψ. Then the stochastic system throughput optimization problem is given as

$$\max_{\pi} E_{\mathbf{S} \sim \psi}[\Phi_{\pi}(\mathbf{S})], \qquad (2.23)$$

where $\Phi_{\pi}(\mathbf{S})$ denotes the system throughput under the channel states $\mathbf{S}$, and $E_{\mathbf{S} \sim \psi}[\cdot]$ denotes the expected system throughput under the channel state distribution ψ. Recent result in [12] shows that the MRAS algorithm can be used to solve such stochastic optimization problem by drawing a large samples of channel-states $\{\mathbf{S}(1), \ldots, \mathbf{S}(L)\}$ from the probability distribution ψ and evaluating the expected performance by the sample average (i.e., $E_{\mathbf{S} \sim \psi}[\Phi_{\pi}(\mathbf{S})] = \frac{1}{L} \sum_{l=1}^{L} \Phi_{\pi}(\mathbf{S}(l))$). When the size of channel-states samples is large enough, the MRAS algorithm can converge to the optimal solution π^* approximately [12]. Based on the idea above, secondary users can first probe the channel environment by sensing and recording the channel states $\{\mathbf{S}(t)\}_{t=1}^{T}$ over a long time period consisting of T time slots. Note that the channel probing can be achieved in a collaborative way that each user selects one channel to sense, and shares the sensing results with other users at end of the probe period. Then each user can apply the MRAS algorithm to compute the near-optimal channel recommendation policy π^* by constituting Φ_{π} as $\frac{1}{T} \sum_{t=1}^{T} \Phi_{\pi}(\mathbf{S}(t))$ in Algorithm 2.

Note that the optimization problem in (2.23) can also be generalized to take other dynamic factors into account. For example, let $\boldsymbol{\varrho} = (\varrho_1, \ldots, \varrho_M)$ denote the loss rates of all the channels, which follow a probability distribution ϕ. Then the stochastic system throughput optimization problem can be written as

$$\max_{\pi} E_{\mathbf{S}\sim\psi,\varrho\sim\phi}[\Phi_{\pi}(\mathbf{S},\varrho)], \tag{2.24}$$

where $\Phi_{\pi}(\mathbf{S},\varrho)$ denotes the expected system throughput under the channel states $\mathbf{S}$ and channel loss rates ϱ. We can solve the problem (2.24) with a similar procedure as described above.

As another example, we can apply the optimization formulation in (2.23) to address the issue of heterogeneous user capacities. Let $\mathbf{a}(t) = (a_1(t),\ldots,a_N(t))$ be the channel selections of all users at time slot t, and let B_n^m denote the mean data rate that user n achieves on channel m. Then the stochastic system throughput optimization problem in (2.23) can be written as

$$\max_{\pi} E_{\mathbf{S}\sim\psi}[\Phi_{\pi}(\mathbf{S})] = \max_{\pi} \frac{1}{T}\sum_{t=1}^{T}\Phi_{\pi}(\mathbf{S}(t))$$

$$= \max_{\pi} E_{\{\mathbf{a}(t)\}_{t=1}^{T}\sim\pi}\left[\frac{1}{T}\sum_{t=1}^{T}U(\mathbf{S}(t),\mathbf{a}(t))\right].$$

where $U(\mathbf{S}(t),\mathbf{a}(t))$ denotes the system throughput under channel states $\mathbf{S}$ and channel selections $\mathbf{a}$, which can be computed as $U(\mathbf{S}(t),\mathbf{a}(t)) = \sum_{n=1}^{N} S_{a_n(t)}(t) B_n^{a_n(t)} \times g_n^{a_n(t)}(\mathbf{a}(t))$. Here $g_n^{a_n}(\mathbf{a})$ denotes the probability that user n successfully grabs the channel a_n, which can be derived from the adopted channel contention mechanism. For the random backoff mechanism in this chapter, we have $g_n^{a_n(t)}(\mathbf{a}(t)) = \frac{1}{\sum_{i=1}^{N} I_{\{a_i(t)=a_n(t)\}}}$. Similarly, by the sample average approach (i.e., drawing L samples of actions over T time slots $\{\mathbf{a}(t)\}_{t=1}^{T}$ from the policy π), we can obtain the expected system throughput as

$$E_{\mathbf{S}\sim\psi}[\Phi_{\pi}(\mathbf{S})] = \frac{\sum_{t=1}^{T}\sum_{l=1}^{L}U(\mathbf{S}(t),\mathbf{a}_l(t))}{TL},$$

and then apply the MRAS algorithm to find the solution.

2.8 Simulation Results

In this section, we investigate the proposed adaptive channel recommendation scheme by simulations. The results show that the adaptive channel recommendation scheme not only achieves a higher performance over the static scheme and random access scheme, but also is more robust to the dynamic change of the channel environments.

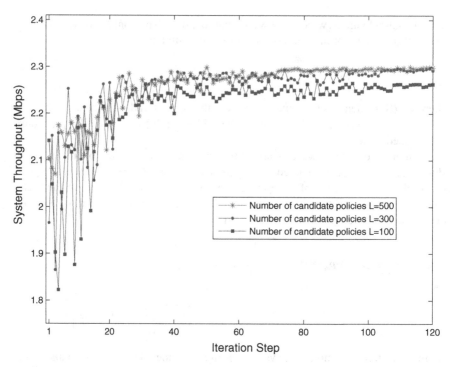

Fig. 2.5 The convergence of MRAS-based algorithm with different number of candidate policies per iteration

2.8.1 Simulation Setup

We initialize the parameters of MRAS algorithm as follows. We set $\mu_R = 0.5$ and $\sigma_R = 0.5$ for the Gaussian distribution, which has 68.2% support over the feasible region $(0, 1)$. We found that the performance of the MRAS algorithm is insensitive to the elite ratio ρ when $\rho \leq 0.3$. We thus choose $\rho = 0.1$.

When using the MRAS-based algorithm, we need to determine how many (feasible) candidate policies to generate in each iteration. Figure 2.5 shows the convergence of MRAS algorithm with 100, 300, and 500 candidate policies per iteration, respectively. We have two observations. First, the number of iterations to achieve convergence reduces as the number of candidate policies increases. Second, the convergence speed is insignificant when the number changes from 300 to 500. We thus choose $L = 500$ for the experiments in the sequel.

Homogeneous Channel Recommendation

We first consider a cognitive radio network consisting of $M = 10$ stochastically homogeneous primary channels, and $N = 5$ secondary users. The data rate of each channel is normalized to be 1 Mbps. In order to take the impact of primary user's long run behavior into account, we consider the following two types of homogeneous channel environments (i.e., channel state transition matrices):

$$\text{Type 1: } \Gamma^1 = \begin{bmatrix} 1 - 0.005\varepsilon & 0.005\varepsilon \\ 0.025\varepsilon & 1 - 0.025\varepsilon \end{bmatrix}, \tag{2.25}$$

$$\text{Type 2: } \Gamma^2 = \begin{bmatrix} 1 - 0.01\varepsilon & 0.01\varepsilon \\ 0.01\varepsilon & 1 - 0.01\varepsilon \end{bmatrix}, \tag{2.26}$$

where ε is the dynamic factor. Recall that a larger ε means that the channels are more dynamic over time. Using (2.2), we know that channel environments Γ^1 and Γ^2 have the stationary channel idle probabilities of $1/6$ and $1/2$, respectively. In other words, the primary activity level is much higher with the Type 1 channel environment than with the Type 2 channel environment. We implement the adaptive channel recommendation scheme, and benchmark it with the static channel recommendation scheme in [5] and the random access scheme. We choose the dynamic factor ε within a wide range to investigate the robustness of the schemes to the channel dynamics. The results are shown in Figs. 2.6 and 2.7. From these figures, we see that the adaptive channel recommendation scheme offers 5–18 % performance gain over the static scheme. Moreover, the adaptive channel recommendation is much more robust to

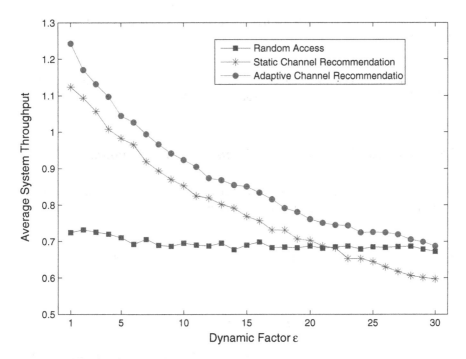

Fig. 2.6 System throughput with $M = 10$ channels and $N = 5$ users under the Type 1 channel state transition matrix

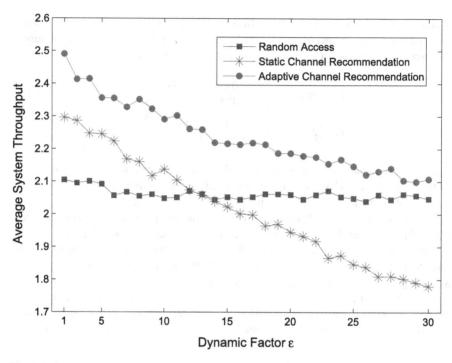

Fig. 2.7 System throughput with $M = 10$ channels and $N = 5$ users under the Type 2 channel state transition matrix

the dynamic channel environment changing. The reason is that the optimal adaptive policy takes the channel dynamics into account while the static one does not.

2.8.2 Heuristic Heterogenous Channel Recommendation

We now evaluate the proposed heuristic heterogeneous channel recommendation mechanism in Sect. 2.6. e implement the heuristic heterogeneous channel recommendation mechanism in heterogenous channel environments. The data rates of $M = 10$ channels are $\{B_1 = 0.2, B_2 = 0.6, B_3 = 0.8, B_4 = 1, B_5 = 2, B_6 = 4, B_7 = 6, B_8 = 8, B_9 = 10, B_{10} = 20\}$ Mbps. The stochastic channel state changing environment is given as:

$$\{\Gamma_1 = \Gamma^1, \Gamma_2 = \Gamma^1, \Gamma_3 = \Gamma^1, \Gamma_4 = \Gamma^1, \Gamma_5 = \Gamma^1,$$
$$\Gamma_6 = \Gamma^2, \Gamma_7 = \Gamma^2, \Gamma_8 = \Gamma^2, \Gamma_9 = \Gamma^2, \Gamma_{10} = \Gamma^2\}. \tag{2.27}$$

Here subscript denotes channel index, and superscript denote channel type index. We also implement static channel recommendation, the optimal homogeneous channel

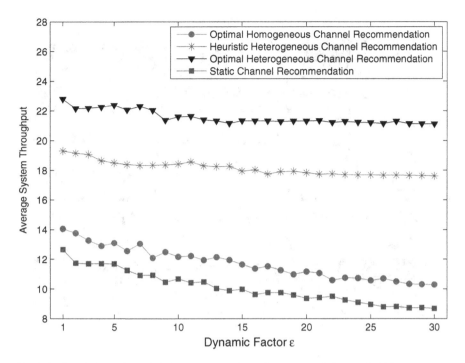

Fig. 2.8 Comparison of heuristic heterogenous channel recommendation, optimal homogeneous channel recommendation and optimal homogeneous channel recommendation

recommendation (Algorithm 1) and optimal heterogeneous channel recommendation (similar with Algorithm 1 by replacing system state R and decision variables P_{rec} with $\mathbf{R}$ and $\{P_m(\mathbf{R})\}_{m=1}^{M}$, respectively) as benchmarks. The results are depicted in Fig. 2.8. From the figure, we see that the heuristic heterogeneous channel recommendation achieves up-to 70 and 100 % performance improvement over the optimal homogeneous channel recommendation and static channel recommendation, respectively. The performance loss is at most 20 % comparing with the the optimal heterogeneous channel recommendation. Note that the number of decision variables in the optimal heterogeneous channel recommendation is $M2^M = 10,240$, while the number of decision variables in the heuristic heterogeneous channel recommendation is only $2M = 20$. The convergence of the heuristic heterogeneous channel recommendation hence is much faster than the optimal heterogeneous channel recommendation.

2.8.3 Simulation with Real Channel Data

We now evaluate the adaptive channel recommendation scheme using real channel data. The data we used (from Xu et al. [13]) is the spectral measurements taken in 850–870 MHz public safety band in Maryland. The measured band is divided

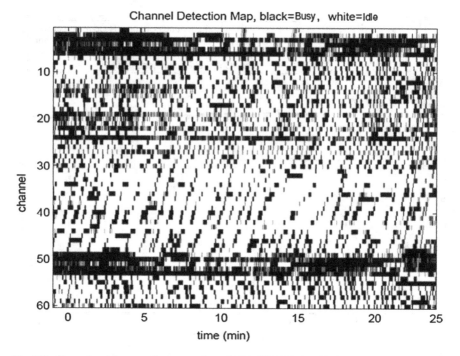

Fig. 2.9 Channel activity map from trace data of 850–870 MHz band in Maryland [13]

into 60 channels, and each channel has a bandwidth of 25 KHz. The measurements were taken over a duration of 25 min, with each time slot being 0.01 s. PU's activity is determined by the energy detection with a threshold of 10 dB above the noise floor [14]. Figure 2.9 visualizes the real trace data. We observe that these channels exhibit a large number of busy/idle cycles (i.e., temporal correlations) and statistically heterogeneous channel availabilities.

We implement the heuristic heterogeneous channel recommendation scheme in a network consisting of 6 channels from the real data. We set the mean data rates of all channels as $\{B_1 = 5, B_2 = 8, B_3 = 12, B_4 = 15, B_5 = 18, B_6 = 20\}$ Mbps. For the channel contention, we set the number of backoff mini-slots $\lambda^* = 20$. Besides the system-wide throughput, we also consider the average access delay, i.e., the average number of time slots that a secondary user needs to wait until its data packet can successfully go through for transmission without blocking. A data packet can be blocked due to the factors such as the channel availability and channel contentions. As a benchmark, we also implement a belief-based channel access scheme proposed in previous work [15, 16] as follows:

- Each user n maintains the following two vectors: $X_n = (X_1^n, \ldots, X_M^n)$ and $Y_n = (Y_1^n, \ldots, Y_M^n)$, where X_m^n and Y_m^n record the number of time slots in which the channel m has been sensed to be free, and the number of time slots in which the channel m has been sensed. Set $X_m^n = Y_m^n = 1$ initially.

- At the beginning of each time slot, each user n computes its belief as $\omega_m^n = \frac{X_m^n}{Y_m^n}$ and chooses each channel m with probability $\frac{\omega_m^n}{\sum_{m'=1}^M \omega_{m'}^n}$.

- At the end of each time slot, each user n broadcasts the sensing result to other users, and then updates the parameters X_n and Y_n based on the overall sensing results of all users.

The key idea of the belief-based channel access is to select channels based on the belief ω_m^n generated from the history of users' observations X_n and Y_n. We implement the adaptive channel recommendation and belief-based channel access schemes with the number of users N ranging from 2 to 8. The results are shown in Figs. 2.10 and 2.11. Compared with the belief-based channel access scheme, we see that the channel recommendation scheme can achieve up-to 30 % system throughput improvement, and reduce up-to 15 % access delay.

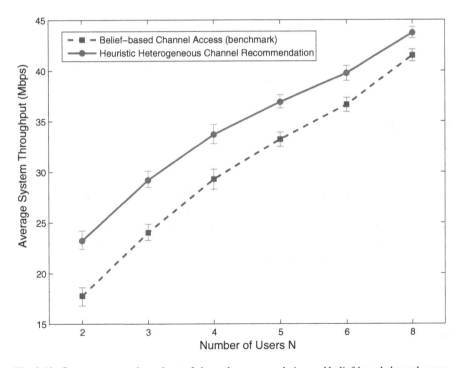

Fig. 2.10 System average throughput of channel recommendation and belief-based channel access (benchmark) schemes with heterogeneous channel date rates

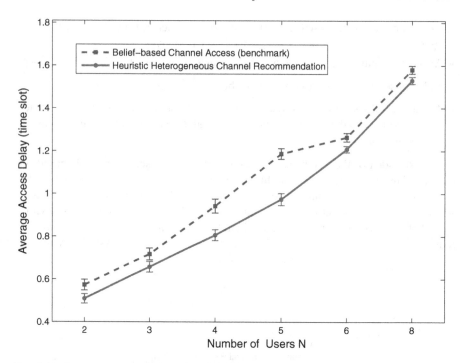

Fig. 2.11 Average access delay of channel recommendation and belief-based channel access (benchmark) schemes with heterogeneous channel date rates

2.9 Summary

In this chapter, we propose an adaptive channel recommendation scheme for efficient spectrum sharing. We formulate the problem as an average reward based Markov decision process. We first prove the existence of the optimal stationary spectrum access policy, and then characterize the structure of the optimal policy in two asymptotic cases. Furthermore, we propose a novel MRAS-based algorithm that is provably convergent to the optimal policy. Numerical results show that our proposed algorithm outperforms the static approach in the literature by up to 100 % in terms of system throughput. Our algorithm is also more robust to the channel dynamics compared to the static counterpart.

References

1. J. Mitola, Cognitive radio: an integrated agent architecture for software defined radio, Ph.D. dissertation, Royal Institute of Technology (KTH) Stockholm, Sweden, 2000
2. Q. Zhao, L. Tong, A. Swami, Y. Chen, Decentralized cognitive MAC for opportunistic spectrum access in ad hoc networks: a pomdp framework. IEEE J. Sel. Areas Commun. **25**, 589–600 (2007)

3. M. Wellens, J. Riihijarvi, P. Mahonen, Empirical time and frequency domain models of spectrum use. Elsevier Phys. Commun. **2**, 10–32 (2009)

4. M. Wellens, J. Riihijarvi, M. Gordziel, P. Mahonen, Spatial statistics of spectrum usage: from measurements to spectrum models, in *IEEE International Conference on Communications* (2009)

5. H. Li, Customer reviews in spectrum: recommendation system in cognitive radio networks, in *IEEE Symposia on New Frontiers in Dynamic Spectrum Access Networks (DySPAN)* (2010)

6. R. Murty, R. Chandra, T. Moscibroda, P. Bahl, Senseless: a database-driven white spaces network, in *IEEE Symposium on New Frontiers in Dynamic Spectrum Access Networks (DySPAN)* (2011)

7. J. Hu, M. Fu, S. Marcus, A model reference adaptive search algorithm for global optimization. Oper. Res. **55**, 549–568 (2007)

8. X. Chen, J. Huang, H. Li, Adaptive channel recommendation for opportunistic spectrum access. IEEE Trans. Mob. Comput. **12**(9), 1788–1800 (2013). http://arxiv.org/pdf/1102.4728.pdf

9. C. Cormio, K.R. Chowdhury, Common control channel design for cognitive radio wireless ad hoc networks using adaptive frequency hopping. Elsevier J. Ad Hoc Netw. **8**, 430–438 (2010)

10. S.M. Ross, *Introduction to Stochastic Dynamic Programming* (Academic Press, New York, 1993)

11. E. Gilbert et al., Capacity of a burst-noise channel. Bell Syst. Tech. J. **39**(9), 1253–1265 (1960)

12. J. Hu, M. Fu, S. Marcus, A model reference adaptive search method for stochastic global optimization. Commun. Inf. Syst. **8**(3), 245–276 (2008)

13. D. Xu, E. Jung, X. Liu, Optimal bandwidth selection in multi-channel cognitive radio networks: how much is too much? in *3rd IEEE Symposium on New Frontiers in Dynamic Spectrum Access Networks (DySPAN)* (2008)

14. S. Jones, E. Jung, X. Liu, N. Merheb, I. Wang, Characterization of spectrum activities in the US public safety band for opportunistic spectrum access, in *IEEE International Symposium on New Frontiers in Dynamic Spectrum Access Networks (DySPAN)* (2007)

15. H. Liu, B. Krishnamachari, Q. Zhao, Cooperation and learning in multiuser opportunistic spectrum access, in *IEEE International Conference on Communications (ICC)* (2008)

16. L. Lai, H. El Gamal, H. Jiang, H. Poor, Cognitive medium access: exploration, exploitation, and competition. IEEE Trans. Mob. Comput. **10**(2), 239–253 (2011)

Chapter 3
Imitative Spectrum Access Mechanism

3.1 Introduction

In this chapter, we will design distributed spectrum access mechanism based on *imitation*, which is also a common phenomenon in many social animal and human interactions [1]. Imitation is simple (just follow a successful action of another user) and turns out to be an efficient strategy in many applications. For example, Schlag [2] used imitation to solve the multi-armed bandit problem. Lopes et al. [3] designed an efficient imitation-based social learning mechanism for robots. Imitation in wireless networks, however, has several fundamental differences from the previous approach. For example, when multiple users imitate the same channel choice, then the data rate obtained by each user will be reduced due to the congestion on that channel.

Recently, Iellamo et al. [4] proposed an imitation-based spectrum access mechanism for spectrum sharing networks, by assuming that all the secondary users are homogeneous (i.e., they experience the same channel condition) and the true expected throughput on a channel is known by a secondary user once the user has chosen the channel. In this chapter, we relax these restrictive assumptions and design an imitative spectrum access mechanism based on user's *local observations* such as the realized data rates and transmission collisions. The key idea is that each user applies the maximum likelihood estimation to estimate its expected throughput, and imitates another neighboring user's channel selection if neighbor's estimated throughput is higher. Moreover, as imitation requires limited information sharing, we introduce the information sharing graph to model the social information sharing relationship among the secondary users. For example, in practical wireless systems it is often the case that a user can only receive message broadcasting from a subset of users that are close enough due to the geographical constraint. Moreover, we also generalize the proposed imitation based spectrum access mechanism to the case that secondary users are heterogeneous. The main results and contributions of this chapter are as follows:

© The Author(s) 2015
X. Chen and J. Huang, *Social Cognitive Radio Networks*,
SpringerBriefs in Electrical and Computer Engineering,
DOI 10.1007/978-3-319-15215-8_3

- *Imitative Spectrum Access*: We propose a novel imitation-based distributed spectrum access mechanism on a general information sharing graph. Each secondary user first estimates its expected throughput based on local observations, and then chooses to imitate a better neighbor. The imitation-based mechanism leverages the social intelligence of the secondary user crowd and only requires a low computational power for each individual user.
- *Convergence to the Imitation Equilibrium*: We show that the imitative spectrum access mechanism converges to the imitation equilibrium, wherein no imitation can be further carried out on the time average. When the information sharing graph is connected, we show that the imitation equilibrium corresponds to a fair channel allocation, such that all the users achieve the same throughput in the asymptotic case.
- *Imitative Spectrum Access with User Heterogeneity*: We further design an imitation-based spectrum access mechanism with user heterogeneity, where different users achieve different data rates on the same channel. Numerical results show that the proposed mechanism achieves up-to 530 % fairness improvement with at most 20 % performance loss, compared with the centralized optimal solution. This demonstrates that the proposed imitation-based mechanism can achieve efficient spectrum utilization and meanwhile provide good fairness across secondary users.

The rest of the chapter is organized as follows. We introduce the system model in Sect. 3.2. We then present the imitative spectrum access mechanism in Sect. 3.3, and study the dynamics and convergence of the imitative spectrum access mechanism in Sect. 3.4. We proposed imitative spectrum access mechanism with user heterogeneity, and illustrate the performance of the proposed mechanisms through numerical results in Sects. 3.5 and 3.6, respectively, Finally, we conclude the chapter in Sect. 3.7. *Due to space limitations, the details for several proofs are provided in* [5].

3.2 System Model

In this part, we first discuss the system model of distributed spectrum sharing, and then introduce the information sharing graph for the imitation mechanism.

3.2.1 Spectrum Sharing System Model

We consider a spectrum sharing network with a set $\mathcal{M} = \{1, 2, \ldots, M\}$ of independent and *stochastically heterogeneous* licensed channels. A set $\mathcal{N} = \{1, 2, \ldots, N\}$ of secondary users try to opportunistically access these channels, when the channels are not occupied by primary (licensed) transmissions. For simplicity, we assume that all secondary users accessing the same channel will interfere with each other (i.e., the interference graph under the protocol interference model [6] is fully meshed).

The case with the spatial reuse (i.e., the interference graph can be partially meshed) will be considered in a future work. The system model has a slotted transmission structure as in Fig. 3.1 and is described as follows.

(1) *Channel State*: the channel state for a channel m during a time slot τ is

$$S_m(\tau) = \begin{cases} 0, & \text{if channel } m \text{ is occupied by primary transmissions,} \\ 1, & \text{if channel } m \text{ is idle.} \end{cases}$$

(2) *Channel State Changing*: for a channel m, we assume that the channel state is an i.i.d. Bernoulli random variable, with an idle probability $\theta_m \in (0, 1)$ and a busy probability $1 - \theta_m$. This model can be a good approximation of the reality if the time slots for secondary transmissions are sufficiently long or the primary transmissions are highly bursty [2]. The motivation of considering the i.i.d. channel state model is to focus our analysis on the spectrum contention due to secondary users' dynamic channel selections. However, numerical results show that the proposed mechanism also works well in the Markovian channel environment where channel states have correlations between time slots. Please refer to Sect. 3.6.1.2 for a detailed discussion.

(3) *Heterogeneous Channel Throughput*: if a channel m is idle, the achievable data rate by a secondary user in each time slot $b_m(\tau)$ evolves according to an i.i.d. random process with a mean B_m, due to the local environmental effects such as fading [3]. For example, we can compute the data rate $b_m(\tau)$ according to the Shannon capacity as

$$b_m(\tau) = E_m \log_2\left(1 + \frac{\eta_n h_m(\tau)}{\upsilon_m}\right), \tag{3.1}$$

where E_m is the bandwidth of channel m, η_n is the fixed transmission power adopted by user n according to the requirements such as the primary user protection, υ_m denotes the background noise power, and $h_m(\tau)$ is the channel gain. In a Rayleigh fading channel environment, the channel gain $h_m(\tau)$ is a random variable that follows the exponential distribution [3]. Here we first consider the homogeneous user case that all users achieve the same mean data rate on the same channel (but users can achieve different data rates on different channels). In Sect. 3.5, we will further consider the heterogeneous user case that different users can achieve different mean data rates even on the same channel. This will allow users to have different transmission technologies, choose different coding/modulation schemes, and experience different channel conditions.

(4) *Time Slot Structure*: each secondary user n executes the following stages synchronously during each time slot:

- *Channel Sensing*: sense one of the channels based on the channel selection decision generated at the end of previous time slot (see below). Access the channel if it is idle.

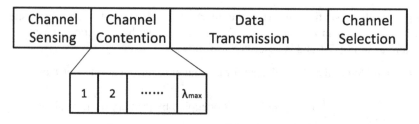

Fig. 3.1 Multiple stages in a single time slot

- *Channel Contention*: use a backoff mechanism to resolve collisions when multiple secondary users access the same idle channel.[1] The contention stage of a time slot is divided into $\lambda_{\max}$ mini-slots[2] (see Fig. 3.1), and user n executes the following two steps. *First*, count down according to a randomly and uniformly chosen integral backoff time (number of mini-slots) λ_n between 1 and $\lambda_{\max}$. *Second*, once the timer expires, transmit RTS/CTS messages if the channel is clear (i.e., no ongoing transmission). Note that if multiple users choose the same backoff value λ_n, a collision will occur with RTS/CTS transmissions and no users win the channel contention.
- *Data Transmission*: transmit data packets if the RTS/CTS message exchange is successful (i.e., the user wins the channel contention).
- *Channel Selection*: choose a channel to access in the next time slot according to the imitative spectrum access mechanism (introduced in Sect. 3.3).

Suppose that k_m users choose the same idle channel m to access. Then the probability that a user n (out of the k_m users) successfully grabs the channel m is

$$g(k_m) = Pr\{\lambda_n < \min_{i \neq n}\{\lambda_i\}\}$$

$$= \sum_{\lambda=1}^{\lambda_{\max}} Pr\{\lambda_n = \lambda\} Pr\{\lambda < \min_{i \neq n}\{\lambda_i\}|\lambda_n = \lambda\}$$

$$= \sum_{\lambda=1}^{\lambda_{\max}} \frac{1}{\lambda_{\max}} \left(\frac{\lambda_{\max} - \lambda}{\lambda_{\max}}\right)^{k_m - 1},$$

which is a decreasing function of the total contending users k_m. Then the long-run expected throughput of a secondary user n choosing a channel m is given as

$$U_n = \theta_m B_m g(k_m). \tag{3.2}$$

---

[1] For ease of exposition, we adopt the backoff mechanism as an example. Our analysis can apply to many other medium access control (MAC) schemes such as TDMA.

[2] Note that in general the length of a mini-slot is much smaller than the length of spectrum sensing and access period in a time slot. For example, for IEEE 802.11af systems (also known as WhiteFi Networks), the length of a mini-slot is 4 microseconds and the spectrum sensing duration is 0.5 milliseconds [7].

3.2.2 Social Information Sharing Graph

In order to carry out imitations, we assume that there exists a common control channel for the information exchange among secondary users.[3] As an alternative, we can adopt the proximity-based communication approach [8], such that secondary users equipped with the radio interfaces such as near field communication (NFC)/bluetooth/WiFi Direct can communicate with each other directly for information exchange. Since information exchange typically would incur an overhead such as the extra energy consumption, it is important to design a proper incentive mechanism for stimulating collaborative information exchange among secondary users. One possible approach is to design mechanisms such that users receive exogenous incentives for cooperation. For example, a payment based incentive mechanism [9] compensates users' contributions by rewarding them with virtual currency. In reputation based incentive mechanisms [10], users' cooperative behaviors are monitored by some centralized authority or collectively by the whole user population, so that any user's selfish behaviors would be detected and punished. In general, such an approach requires centralized infrastructures (e.g., secondary base-station/access point), which would incur a high system overhead and may not be feasible in our context of distributed spectrum sharing.

The centralized infrastructures are not available, motivated by the observation that the hand-held devices are typically carried by human beings, we can leverage the endogenous incentive which comes from the intrinsic social relationships among users to promote effective and trustworthy cooperation. For example, when a user is at home or work, typically family members, neighbors, colleagues, or friends are nearby. The user can then exploit the social trust from these neighboring users to achieve effective cooperation for information exchange. Indeed, with the explosive growth of online social networks such as Facebook and Twitter, more and more people are actively involved in online social interactions, and social connections among people are being extensively broadened. This has opened up a new avenue to integrate the social interactions for cooperative networking design.

Specifically, we introduce the social information sharing graph $\mathcal{G} = \{\mathcal{N}, \mathcal{E}\}$ to model cooperative information exchange relationships due to the social ties among the secondary users. Here the vertex set is the same as the user set $\mathcal{N}$, and the edge set is given as $\mathcal{E} = \{(n, m) : e_{nm} = 1, \forall n, m \in \mathcal{N}\}$ where $e_{nm} = 1$ if and only if users n and m have social tie between each other, e.g., kinship, friendship, or colleague relationships. Furthermore, for a pair of users n and m who have a social edge between them on the social graph, we formalize the strength of social tie as $\delta_{nm} \in [0, 1]$, with a higher value of δ_{nm} being a stronger social tie. Each secondary user n can specify a cooperation threshold φ_n and is willing to share information

---

[3] There are several approaches for establishing a common control channel in cognitive radio networks, e.g., sequence-based rendezvous [11], adaptive channel hopping [12] and user grouping [13]. Please refer to [14] for a comprehensive survey on the research of common control channel establishment in cognitive radio networks.

with those users with whom he has a high enough social tie above the cooperation threshold φ_n. Moreover, to thwart the potential attacks of releasing false channel information by malicious users and enhance the security level of imitation based spectrum access, each secondary user n can set a trust threshold η_n and choose to trust the information from those users having a high enough social tie above the trust threshold η_n. In the sequel, we denote the neighborhood of user n for effective and trustworthy information sharing as $\mathcal{N}_n \triangleq \{k : e_{nk} = 1 \text{ and } \delta_{nk} \geq \eta_n \text{ and } \delta_{kn} \geq \varphi_k\}$. In terms of implementation, the social relationship identification procedure can be carried out prior to the imitative spectrum access. Specifically, two secondary users can locally initiate the "matching" process to detect the common social features between them. For example, two users can match their contact lists. If they have the phone numbers of each other or many of their phone numbers are the same, then it is very likely that they know each other. As another example, two device users can match their home and working addresses and identify whether they are neighbors or colleagues. To preserve the privacy of the secondary users, the private set intersection and homomorphic encryption techniques proposed in [15, 16] can be adopted to design a privacy-preserving social relationship identification mechanism.

We should emphasize that when it is difficult to leverage the social trust among some secondary users and the centralized infrastructures are not available, similar to the file sharing in the P2P systems [17], we can adopt the Tit-for-Tat mechanism for information sharing. Specifically, based on the principle of reciprocity, a secondary user will always share information with its partner as long as its partner (i.e., another user) also shares information with it. If the partner refuses to share information, the user will punish its partner by not sharing information either. As a result, the partner would suffer and learn to share information with the user again. Notice that since the imitative spectrum access mechanism can work on a generic social information sharing graph, we can also use a hybrid approach of several different schemes mentioned above for establishing the information sharing relationships among the secondary users.

Since our analysis is from secondary users' perspective, we will use terms "secondary user" and "user" interchangeably in the following sections.

3.3 Imitative Spectrum Access Mechanism

We now apply the idea of imitation to design an efficient distributed spectrum access mechanism, which utilizes user's local estimation of its expected throughput. Each user randomly chooses a neighboring user in the information sharing graph, and follows the neighbor's channel selection if the neighbor's throughout is better than its.

3.3.1 Expected Throughput Estimation

In order to imitate a successful action, a user needs to compare its and other users' performances (throughputs). In practice, many wireless devices only have a limited view of the network environment due to hardware constraints. To incorporate the effect of incomplete network information, we first introduce the maximum likelihood estimation (MLE) approach to estimate user's expected throughput based on its local observations. We choose MLE mainly due to the efficiency and the ease of implementation of this method [18]. To achieve accurate local estimation based on local observations, a user needs to gather a large number of observation samples. This motivates us to divide the spectrum access time into a sequence of *decision periods* indexed by t ($t = 1, 2, \ldots$), where each decision period consists of L time slots (see Fig. 3.2 for an illustration). During a single decision period, a user accesses the *same* channel in all L time slots. Thus the total number of users accessing each channel does not change within a decision period, which allows users to learn the environment.

According to (4.1), a user's expected throughput during decision period t depends on the probability of grabbing the channel $g(k_m(t))$ on that period, the channel idle probability θ_m, and the mean data rate B_m.

3.3.1.1 MLE of Channel Grabbing Probability $g(k_m(t))$

At the beginning of each time slot τ ($\tau = 1, \ldots, L$) of a decision period t, we assume that a user n chooses to sense the same channel m. If the channel is idle, the user will contend to grab the channel according to the backoff mechanism in Sect. 3.2. At the end of each time slot τ, a user observes $S_n(t, \tau)$, $I_n(t, \tau)$, and $b_n(t, \tau)$. Here $S_n(t, \tau)$ denotes the state of the chosen channel (i.e., whether occupied by the primary traffic), $I_n(t, \tau)$ indicates whether the user has successfully grabbed the channel, i.e.,

$$I_n(t, \tau) = \begin{cases} 1, & \text{if user } n \text{ successfully grabs the channel} \\ 0, & \text{otherwise,} \end{cases}$$

and $b_n(t, \tau)$ is the received data rate on the chosen channel by user n at time slot τ. Note that if $S_n(t, \tau) = 0$ (i.e., the channel is occupied by the primary traffic), we set $I_n(t, \tau)$ and $b_n(t, \tau)$ to be 0. At the end of each decision period t, each user n will have a set of local observations $\Omega_n(t) = \{S_n(t, \tau), I_n(t, \tau), b_n(t, \tau)\}_{\tau=1}^{L}$.

When channel m is idle (i.e., no primary traffic), consider $k_m(t)$ users contend for the channel according to the backoff mechanism in Sect. 3.2. Then a particular user n

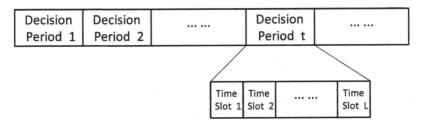

Fig. 3.2 The period structure of maximum likelihood estimation of various system parameters

out of these $k_m(t)$ users grabs the channel with the probability $g(k_m(t))$. Since there are a total of $\sum_{\tau=1}^{L} S_n(t, \tau)$ rounds of channel contentions in the period t and each round is independent, the total number of successful channel captures $\sum_{\tau=1}^{L} I_n(t, \tau)$ by user n follows the Binomial distribution. User n then computes the likelihood of $g(k_m(t))$, i.e., the probability of the realized observations $\Omega_n(t)$ given the parameter $g(k(t))$ as

$$\mathscr{L}[\Omega_n(t)|g(k_m(t))] = \begin{pmatrix} \sum_{l=1}^{L} S_n(t, l) \\ \sum_{l=1}^{L} I_n(t, l) \end{pmatrix} g(k_m(t))^{\sum_{l=1}^{L} I_n(t,l)}$$

$$\times (1 - g(k_m(t)))^{\sum_{l=1}^{L} S_n(t,l) - \sum_{l=1}^{L} I_n(t,l)}.$$

Then MLE of $g(k_m(t))$ can be computed by maximizing the log-likelihood function $\ln \mathscr{L}[\Omega_n(t)|g(k_m(t))]$, i.e., $\max_{g(k_m(t))} \ln \mathscr{L}[\Omega_n(t)|g(k_m(t))]$. By the first order condition, we obtain the optimal solution as $\tilde{g}(k_m(t)) = \sum_{\tau=1}^{L} I_n(t, \tau)/\sum_{\tau=1}^{L} S_n(t, \tau)$, which is the sample averaging estimation. When the length of decision period L is large, by the central limit theorem, we know that $\tilde{g}(k_m(t)) \sim \mathscr{N}\left(g(k_m(t)), \frac{g(k_m(t))(1-g(k_m(t)))}{\sum_{\tau=1}^{L} S_n(t,\tau)}\right)$, where $\mathscr{N}(\cdot)$ denotes the normal distribution.

3.3.1.2 MLE of Channel Idle Probability θ_m

We next apply the MLE to estimate the channel idle probability θ_m. Since the channel state $S_n(t, \tau)$ is i.i.d over different time slots and different decision periods, we can improve the estimation by averaging not only over multiple time slots but also over multiple periods.

Similarly with MLE of $g(k_m(t))$, we first compute one-period MLE of θ_m as $\hat{\theta}_m = \frac{\sum_{\tau=1}^{L} S_n(t,\tau)}{L}$. When the length of decision period L is large, we have that $\hat{\theta}_m$ follows the normal distribution with the mean θ_m, i.e., $\hat{\theta}_m \sim \mathscr{N}\left(\theta_m, \frac{\theta_m(1-\theta_m)}{L}\right)$.

We then average the estimation over multiple decision periods. When a user n finishes accessing a channel m for a total C periods, it updates the estimation of the channel idle probability θ_m as $\tilde{\theta}_m(C) = \frac{1}{C} \sum_{i=1}^{C} \hat{\theta}_m(i)$, where $\tilde{\theta}_m(C)$ is the estimation of θ_m based on the information of all C decision periods, and $\hat{\theta}_m(i)$ is the one-period estimation. By doing so, we have $\tilde{\theta}_m(C) \sim \mathscr{N}\left(\theta_m, \frac{\theta_m(1-\theta_m)}{CL}\right)$, which reduces the variance of one-period MLE by a factor of C.

3.3.1.3 MLE of Average Data Rate B_m

Since the received data rate $b_n(t, \tau)$ is also i.i.d over different time slots and different decision periods, similarly with the MLE of the channel idle probability θ_m, we can

---

Algorithm 3 Imitative Spectrum Access

1: **initialization:**
2:    **choose** a channel a_n randomly for each user n.
3: **end initialization**

4: **loop** for each decision period t and each user n in parallel:
5:    **for** each time slot τ in the period t **do**
6:       **sense** and **contend** to access the channel a_n.
7:       **record** the observations $S_n(t, \tau)$, $I_n(t, \tau)$ and $b_n(t, \tau)$.
8:    **end for**
9:    **estimate** the expected throughput $\tilde{U}_n(t)$.
10:   **select** another user $n' \in \mathcal{N}_n$ randomly and **enquiry** its estimated throughput $\tilde{U}_{n'}(t)$.
11:   **if** $\tilde{U}_{n'}(t) > \tilde{U}_n(t)$ **then**
12:      **choose** channel $a_{n'}$ (i.e., the one chosen by user n') in the next period.
13:   **else choose** the original channel in the next period.
14:   **end if**
15: **end loop**

---

obtain the one-period MLE of mean data rate B_m as $\hat{B}_m = \frac{\sum_{\tau=1}^{L} b_n(t,\tau)}{\sum_{\tau=1}^{L} I_n(t,\tau)}$, and the averaged MLE estimation over C periods as $\tilde{B}_m(C) = \frac{1}{C} \sum_{i=1}^{C} \hat{B}_m(i)$.

By the MLE, we can obtain the estimation of $g(k(t))$, θ_m, and B_m as $\tilde{g}(k_m(t))$, $\tilde{\theta}_m$ and $\tilde{B}_m$, respectively, and then estimate the true expected throughput $U_n(t) = \theta_m B_m g(k(t))$ as $\tilde{U}_n(t) = \tilde{\theta}_m \tilde{B}_m \tilde{g}(k_m(t))$. Since $\tilde{g}(k_m(t))$, $\tilde{\theta}_m$, and $\tilde{B}_m$ follow independent normal distributions with the mean $g(k_m(t))$, θ_m, and B_m, respectively, we thus have $E[\tilde{U}_n(t)] = E[\tilde{\theta}_m \tilde{B}_m \tilde{g}(k_m(t))] = U_n(t)$, i.e., the estimation of expected throughput $U_n(t)$ is unbiased. In the following analysis, we hence assume that

$$\tilde{U}_n(t) = U_n(t) + \omega_n, \tag{3.3}$$

where $\omega_n \in (\underline{\omega}, \overline{\omega})$ is the random estimation noise with the probability density function $f(\omega)$ satisfying

$$f(\omega) > 0, \quad \forall \omega \in (\underline{\omega}, \overline{\omega}), \tag{3.4}$$

$$E[\omega_n] = \int_{\underline{\omega}}^{\overline{\omega}} \omega f(\omega) d\omega = 0. \tag{3.5}$$

3.3.2 Imitative Spectrum Access

We now propose the imitative spectrum access mechanism in Algorithm 3. The key motivation is that, by leveraging the social intelligence of the secondary user crowd, the imitation based mechanism only requires a low computational power for each individual user. More specifically, we let users imitate the actions of those neighboring users that achieve a higher throughput (i.e., Lines 11 to 14 in Algorithm 3).

This mechanism only relies on local throughput comparisons and is easy to implement in practice. Each user n at each period t first collects the local observations $\Omega_n(t) = \{S_n(t, \tau), I_n(t, \tau), b_n(t, \tau)\}_{\tau=1}^L$ (i.e., Lines 5 to 8 in Algorithm 3) and estimates its expected throughput with the MLE method as introduced in Sect. 3.3.1 (i.e., Line 9 in Algorithm 3). Then user n carries out the imitation by randomly sampling the estimated throughput of another user who shares information with him (i.e., Line 10 in Algorithm 3). Such a random sampling can be achieved in different ways. For example, user n can randomly generate a user ID n' from the set $\mathcal{N}_n$ and broadcast a throughput enquiry packet including the enquired user ID n'. Then user n' will send back an acknowledgement packet including the estimation of its own expected throughput.

Intuitively, the benefits of adopting the imitation based channel selection are two-fold. On one hand, since each user has incomplete network information, by enquiring another user's throughput information, each user would have a better view of channel environment. If a channel offers a higher data rate, more users trend to exploit the channel due to the nature of imitation. On the other hand, if too many users are utilizing the same channel, a user can improve its data rate through congestion mitigation by imitating users on another channel with less contending users. In the following Sect. 3.4, we show that the proposed imitation-based mechanism can drive a balance between good channel exploitation and congestion mitigation, and achieve a fair spectrum sharing solution.

We shall emphasize that, in the imitative spectrum access mechanism, we require that each user can (randomly) select *only one* user for the throughput enquiry, in order to promote diversity in users' channel selections for further congestion mitigation and reduce the system overhead for information exchange. We also evaluate the imitative spectrum access schemes, such that each user can select multiple users for throughput enquiry and imitate the channel selection of the best user among these inquired users (please refer to [5] for more details). We observe that the performance of the imitative spectrum access decreases as the number of users for throughput enquiry increases. This is because when each user imitates the best channel selection from multiple users, as the number of enquired users increases, the probability that more users will simultaneously select the same good channel to access in next time slot will increase. This would reduce the diversity of users' channel selections (i.e., increases channel congestion) and hence lead to performance degradation in spectrum sharing, compared with the case of randomly enquiring only one user. Moreover, enquiring multiple users in the same time period will incur a higher system overhead for information exchange.

3.4 Convergence of Imitative Spectrum Access

We then investigate the convergence of imitative spectrum access. Since users' imitations reply on the information exchange, the structure of the information sharing graph hence plays an important role on the convergence of the mechanism. To better understand the structure property, we will introduce an equivalent and yet more compact cluster-based graphical representation of the information sharing graph.

3.4.1 Cluster-Based Graphical Representation of Information Sharing Graph

We now introduce the cluster-based presentation. The cluster concept here is similar with the community structure in social networks analysis [19, 20]. Intuitively, a cluster here can be viewed as a set of users who have similar information sharing structure. Formally, we define that

Definition 3.1 (*Cluster*) A set of users form a cluster if they can share information with each other and they can also share information with the same set of users that are out of the cluster.

Taking the information sharing graph on the left hand-side in Fig. 3.3 as an example, we see that users 1 to 4 form a cluster. However, users 1 to 5 do not form a cluster, since user 5 shares information with user 7 while user 1 does not. Furthermore, we can regard a single user as a special case of cluster. In this case, a general information sharing graph can be represented compactly as a cluster-based graph. Let $\mathcal{K} = \{1, 2, \ldots, K\}$ be the set of clusters, and $w_{kk'} \in \{0, 1\}$ denote the information sharing relationship between two clusters k and k'. The variable $w_{kk'} = 1$ if cluster k communicates with cluster k' (i.e., the users in cluster k share information with the users in cluster k') and $w_{kk'} = 0$ otherwise. Then we denote the cluster-based graph as $\mathcal{CG} = \{\mathcal{K}, \mathcal{W}\}$. Here vertex set $\mathcal{K}$ is the cluster set, and

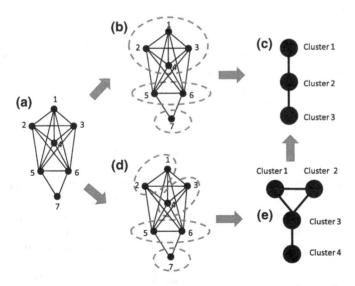

Fig. 3.3 An illustration of cluster-based representation of information sharing graph

---

Algorithm 4 Algorithm for Constructing Cluster-based Graph

1: ▷ **Construct the set of clusters:**
2: **set** the un-merged node set $\mathscr{U} = \mathscr{N}$.
3: **set** cluster index $k = 0$.
4: **loop until** $\mathscr{U} = \varnothing$:
5:    **select** one node $n \in \mathscr{U}$ randomly.
6:    **update** cluster index $k = k + 1$.
7:    **set** the set of nodes in cluster k as $\Omega_k = \{n\}$.
8:    **for** each node $m \in \mathscr{N}_n \cap \mathscr{U} \backslash \Omega_k$ **do**
9:       **if** $\mathscr{N}_n \backslash \{m\} = \mathscr{N}_m \backslash \{n\}$ **then**
10:          **update** $\Omega_k = \Omega_k \cup \{m\}$.
11:       **end if**
12:    **end for**
13:    **update** $\mathscr{U} = \mathscr{U} \backslash \Omega_k$.
14: **end loop**
15: ▷ **Construct the set of edges between clusters:**
16: **set** the set of K identified clusters above as $\Upsilon = \{1, \ldots, K\}$.
17: **for** each cluster $k \in \Upsilon$ **do**
18:    **for** any cluster $h \in \Upsilon \backslash \{k\}$ **do**
19:       **if** there exists nodes $n \in \Omega_k$ and $m \in \Omega_h$ such that $m \in \mathscr{N}_n$ and $n \in \mathscr{N}_m$ **then**
20:          **set** $w_{kh} = 1$.
21:       **else set** $w_{kh} = 0$.
22:       **end if**
23:    **end for**
24: **end for**

---

edge set $\mathscr{W} = \{(k, k') : w_{kk'} = 1, \forall k, k' \in \mathscr{K}\}$. We also denote the set of clusters that communicates with cluster k as $\mathscr{K}_k = \{k : (i, k) \in \mathscr{W}, \forall k \in \mathscr{K}\}$. Since the users in cluster k and the users in cluster $k' \in \mathscr{K}_k$ can share information with each other, we also define that $\mathscr{C}_k \triangleq \mathscr{K}_k \cup \{k\}$.

As illustrated in Fig. 3.3, an information sharing graph can be represented as different cluster-based graphs (e.g., graphs (c) and (e) in Fig. 3.3). In general fact, we can first consider the original information sharing graph as a primitive cluster-based graph by regarding each single user as a cluster (e.g., graph (a) in Fig. 3.3). We then further carry out the clustering (e.g., graph (d) in Fig. 3.3) and obtain the cluster-based graph (e). We next merge clusters 1 and 2 of this cluster-based graph into one cluster and then obtain the most compact cluster-based graph (c) in this example. We summarize the algorithm for constructing the cluster-based graph in Algorithm 4. Note that for the practical implementation, the knowledge of cluster-based graphs is not required. The use of clutter-based graph here is to facilitate the analysis of the convergence of imitative spectrum access mechanism. The convergence properties of the imitative spectrum access mechanism are determined by the original information exchange graph, and are the same for all these cluster-based graphs since they preserve the structural property of the original information exchange graph (i.e., two users share information in the cluster-based graph if and only if they share information in the original graph).

Next we explore the property of the cluster-based representation. We denote the cluster that a user $n \in \mathscr{N}$ belongs to as $k(n)$ and the set of users in cluster $k \in \mathscr{K}$

as $\mathcal{N}(k)$. According to the definition of cluster, we can see that if users n and n' share information with each other, then they either belong to the same cluster or two different clusters that communicate with each other. Thus we have that

Lemma 3.1 *The set of users that share information with user n is the same as the set of users in user n's cluster and the clusters that communicate with user n's cluster, i.e., $\mathcal{N}_n = \cup_{k' \in \mathscr{C}_{k(n)}} \mathcal{N}(k')$.*

Furthermore, the cluster-based representation also preserves the connectivity of the information sharing graph (i.e., it is possible to find a path from any node to any other node).

Lemma 3.2 *The information sharing graph is connected if and only if the corresponding cluster-based graph is also connected.*

3.4.2 Dynamics of Imitative Spectrum Access

Based on the cluster-based graphical representation of information sharing graph, we next study the evolution dynamics of the imitative spectrum access mechanism. Suppose that the underlying information sharing graph can be represented by K clusters, and the number of users in each cluster k is z_k with $\sum_{k=1}^{K} z_k = N$. For the ease of exposition, we will focus the case that the number of users z_k in each cluster k is large. Numerical results show that the observations also hold for the case that the number of users in a cluster is small (see Sects. 3.6 for details).

With a large cluster user population, it is convenient to use the population state $x(t)$ to describe the dynamics of spectrum access. We then denote the population state of all users as $x(t) \triangleq (x^1(t), \ldots, x^K(t))$ and the population state of cluster k as $x^k(t) \triangleq (x_1^k(t), \ldots, x_M^k(t))$. Here $x_m^k(t)$ denotes the fraction of users in cluster k choosing channel m to access at period t, and we have $\sum_{m=1}^{M} x_m^k(t) = 1$.

In the imitative spectrum access mechanism, each user n relies on its local estimated expected throughput $\tilde{U}_n(t)$ to decide whether to imitate other user's channel selection. Due to the random estimation noise ω_n, the evolution of the population state $\{x(t), \forall t \geq 0\}$ is stochastic and difficult to analyze directly. However, when the population of cluster users z_k is large, due to the law of large number, such stochastic process can be well approximated by its mean deterministic trajectory $\{X(t), \forall t \geq 0\}$ [21]. Here $X(t) \triangleq (X^1(t), \ldots, X^K(t))$ is the deterministic population state of all the users, and $X^k(t) \triangleq (X_1^k(t), \ldots, X_M^k(t))$ is the deterministic population state of cluster k. Consider a user in cluster k chooses channel i , and let $P_{i,k}^j(X(t))$ denote the probability that this user in the deterministic population state $X(t)$ will choose channel j in next period. According to [21], we have

Lemma 3.3 *There exists a scalar δ such that, for any bound $\varepsilon > 0$, decision period $T > 0$, and any large enough cluster size z_k, the maximum difference between the*

stochastic and deterministic population states over all periods is upper-bounded by
ε *with an arbitrarily large probability, i.e.,*

$$Pr\{\max_{0\leq t\leq T}\max_{m\in\mathcal{M}}|X_m^k(t) - x_m^k(t)| \geq \varepsilon\} \leq e^{-\varepsilon^2 z_k}, \quad \forall k \in \mathcal{K}, \qquad (3.6)$$

given that $X(0) = x(0)$.

The proof is similar with Lemma 1 in [21] and hence is omitted here. As illustrated in Fig. 3.4, Lemma 3.3 indicates that the trajectory of the stochastic population state $\{x(t), \forall t \geq 0\}$ is within a small neighborhood of the trajectory of the deterministic population state $\{X(t), \forall t \geq 0\}$ when the user population N is large enough. Moreover, since the MLE is unbiased, the deterministic population state $\{X(t), \forall t \geq 0\}$ is also the mean field dynamics of stochastic population state $\{x(t), \forall t \geq 0\}$ [21]. If the deterministic dynamics $\{X(t), \forall t \geq 0\}$ converge to an equilibrium, the stochastic dynamics $\{x(t), \forall t \geq 0\}$ must also converge to the same equilibrium on the time average [21].

We now study the evolution dynamics of the deterministic population state $\{X(t), \forall t \geq 0\}$. Let $U(m, X(t)) = \theta_m B_m g\left(\sum_{k=1}^K z_k X_m^k(t)\right)$ denote the expected throughput of a user that chooses channel m with a total of $\sum_{k=1}^K z_k X_m^k(t)$ contending users in the population state $X(t)$. Recall that in the imitative spectrum access mechanism, each user will randomly choose another user that shares information with it, and imitate that user's channel selection if that user's estimated throughput is higher. Suppose that the user n is in cluster k choosing channel i. According to Lemma 3.1, the set of users that share information with user n are in set of clusters $\mathscr{C}_k$. Thus, we can obtain the probability $P_{i,k}^j(X(t))$ that this user n will imitate another user n' on channel j in next period as

$$P_{i,k}^j(X(t)) = \sum_{k'\in\mathscr{C}_k} \frac{z_{k'}}{\sum_{l\in\mathscr{C}_k} z_l} X_j^{k'}(t)$$
$$\times Pr\{\tilde{U}(j, X(t)) > \tilde{U}(i, X(t))\}. \qquad (3.7)$$

Fig. 3.4 Illustration of the approximation of stochastic population state $x(t)$ by deterministic population state $X(t)$

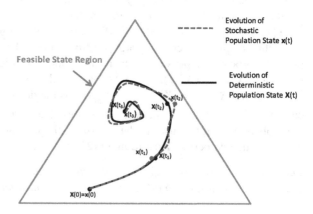

Here $\frac{z_{k'}}{\sum_{l \in \mathscr{C}_k} z_l} X_j^{k'}(t)$ denotes the probability that a user choosing channel j in cluster $k' \in \mathscr{C}_k$ will be selected for imitation. From (3.3), we have

$$
\tilde{U}(j, X(t)) - \tilde{U}(i, X(t)) = U(j, X(t)) - U(i, X(t))
$$
$$
+ \omega_{n'} - \omega_n, \tag{3.8}
$$

where $\omega_n, \omega_{n'}$ are the random estimation noises with the probability density function $f(\omega)$. Let $\varpi = \omega_n - \omega_{n'}$, and we can obtain the probability density function of random variable ϖ as

$$
q(\varpi) = \int_{\underline{\omega}}^{\overline{\omega}} f(\omega) f(\varpi + \omega) d\omega. \tag{3.9}
$$

We further denote the cumulative distribution function ϖ as $Q(\varpi)$, i.e., $Q(\varpi) = \int_{-\infty}^{\varpi} q(s) ds$. Then from (3.7) and (3.8), we have for any $j \neq i$,

$$
P_{i,k}^j(X(t)) = \sum_{k' \in \mathscr{C}_k} \frac{z_{k'}}{\sum_{l \in \mathscr{C}_k} z_l} X_j^{k'}(t) Q(U(j, X(t)) - U(i, X(t))), \tag{3.10}
$$

and

$$
P_{i,k}^i(X(T)) = 1 - \sum_{j \neq i} P_{i,k}^j(X(t)). \tag{3.11}
$$

Based on (3.10) and (3.11), we obtain the evolution dynamics of the deterministic population state $\{X(t), \forall t \geq 0\}$ as

Theorem 3.1 *For the imitative spectrum access mechanism, the evolution dynamics of deterministic population state $\{X(t), \forall t \geq 0\}$ are given as*

$$
\dot{X}_m^k(t) = \sum_{i=1}^M X_i^k(t) \sum_{k' \in \mathscr{C}_k} \frac{z_{k'}}{\sum_{l \in \mathscr{C}_k} z_l} X_m^{k'}(t)
$$
$$
\times Q(U(m, X(t)) - U(i, X(t)))
$$
$$
- X_m^k(t) \sum_{i=1}^M \sum_{k' \in \mathscr{C}_k} \frac{z_{k'}}{\sum_{l \in \mathscr{C}_k} z_l} X_i^{k'}(t)
$$
$$
\times Q(U(i, X(t)) - U(m, X(t))), \tag{3.12}
$$

where the derivative is with respect to time t.

3.4.3 Convergence of Imitative Spectrum Access

We now study the convergence of the imitative spectrum access mechanism. Let $x^* \triangleq (x^{1*}, \dots, x^{K*})$ denote the equilibrium of the imitative spectrum access, and

a_n^* denote the channel chosen by user n in the equilibrium x^*. We first introduce the definition of *imitation equilibrium.*

Definition 3.2 (*Imitation Equilibrium*) A population state x^* is an imitation equilibrium if and only if for each user $n \in \mathcal{N}$,

$$U(a_n^*, x^*) \geq \max_{a \in \Delta_n(x^*) \setminus \{a_n^*\}} U(a, x^*), \tag{3.13}$$

where $\Delta_n(x^*) \triangleq \{m \in \mathcal{M} : \exists a_i^* = m, \forall i \in \mathcal{N}_n\}$ denotes the set of channels are chosen by users that share information with user n in the equilibrium x^*.

The intuition of Definition 3.2 is that no imitation can be carried out to improve any user's data rate in the equilibrium. For the imitative spectrum access mechanism, we show that

Theorem 3.2 *For the imitative spectrum access mechanism, the evolution dynamics of deterministic population state $\{X(t), \forall t \geq 0\}$ asymptotically converge to an imitation equilibrium X^* such that*

$$U(m, X^*) = U(i, X^*), \forall m, i \in \Delta_n(X^*), \quad \forall n \in \mathcal{N}. \tag{3.14}$$

According to Lemma 3.3, we know that the stochastic imitative spectrum access dynamics $\{x(t), \forall t \geq 0\}$ will be attracted into a small neighborhood around the imitation equilibrium X^*. Moreover, since the imitation equilibrium X^* is also the mean field equilibrium of stochastic dynamics $\{x(t), \forall t \geq 0\}$, the stochastic dynamics $\{x(t), \forall t \geq 0\}$ hence converge to the imitation equilibrium X^* on the time average. That is, the fraction of users adopting a certain channel selection will converge to a fixed vale on the time average. However, a user would keep switching its channel during the process. This is because that when many other users also utilize the same channel, the user would imitate to select another channel with less contending users to mitigate congestion. The mechanism hence can drive a balance between good channel exploitation and congestion mitigation.

According to the definition of $\Delta_n(X^*)$, we see from Theorem 3.2 that two users will achieve the same expected throughput if they share information with each other (i.e., they are neighbors in the information sharing graph). Moreover, when the information sharing graph is connected, we can show that all the users achieve the same throughput at the imitation equilibrium.

Lemma 3.4 *When the information sharing graph is connected, all the users following the imitative spectrum access mechanism achieve the same expected throughput, i.e., $U(a_n^*, X^*) = U(a_{n'}^*, X^*), \forall n, n' \in \mathcal{N}$.*

Furthermore, we can show in Corollary 3.5 that the convergent imitation equilibrium is the most fair channel allocation in terms of the widely-used Jain's fairness index $J = \left(\dfrac{(\sum_{n=1}^{N} U(a_n^*, X^*))^2}{N \sum_{n=1}^{N} U(a_n^*, X^*)^2} \right)$ [22]. Notice that the fair channel allocation is due to the nature of imitation. If the channel allocation is unfair, there must exist some secondary users that achieve a higher throughput than others. In this case, other users with a lower throughput will imitate the channel selection of those users until the performance of all users are equal (i.e., fair spectrum sharing).

Lemma 3.5 *When the information sharing graph is connected, the Jain's fairness index J is maximized at the imitation equilibrium.*

We next discuss the efficiency of the imitation equilibrium. Similar to the definition of price of anarchy (PoA) in game theory, we will quantify the efficiency ratio of imitation equilibrium X^* over the centralized optimal solution and define the price of imitation (PoI) as

$$\text{PoI} = \frac{\sum_{n=1}^{N} U(a_n^*, X^*)}{\max_X \sum_{n=1}^{N} U(a_n, X)}.$$

Since it is difficult to analytically characterize the PoI for the general case, we focus on the case that all the channels are homogenous, i.e., $B_m = B_{m'} = B$ and $\theta_m = \theta_{m'} = \theta$ for any $m, m' = 1, \ldots, M$. Let Z be the number of channels being utilized at the imitation equilibrium X^*, i.e., $Z = |\cup_{n=1}^{N} \Delta_n(X^*)|$. We can show the following result.

Theorem 3.3 *When the information sharing graph is connected and all the channels are homogenous, the PoI of imitative spectrum access mechanism is at least $\frac{Ng(\frac{N}{Z})}{M}$.*

We also evaluate the performance of imitative spectrum access mechanism for the general case in Sect. 3.6. Numerical results demonstrate that the mechanism is efficient, with at most 20 % performance loss, compared with the centralized optimal solution.

3.5 Imitative Spectrum Access with User Heterogeneity

For the ease of exposition, we have considered the case that users are homogeneous, i.e., different users achieve the same data rate on the same channel. We now consider the general heterogeneous case where different users may achieve different data rates on the same channel.

Let $b_m^n(\tau)$ be the realized data rate of user n on an idle channel m at a time slot τ, and B_m^n be the mean data rate of user n on the idle channel m, i.e., $B_m^n = E[b_m^n(\tau)]$. In this case, the expected throughput of user n is given as $U_n^m = \theta_m B_m^n g(k_m)$. For imitative spectrum access mechanism in Algorithm 3, each user carries out the channel imitation by comparing its throughput with the throughput of another user. However, such throughput comparison may not be feasible when users are heterogeneous, since a user may achieve a low throughput on a channel that offers a high throughput for another user.

To address this issue, we propose a new imitative spectrum access mechanism with user heterogeneity in Algorithm 5. More specifically, when a user n on a channel

m randomly selects another neighboring user n' on another channel m', user n' informs user n about the estimated channel grabbing probability $\tilde{g}(k_{m'})$ instead of the estimated expected throughput. Then user n will compute the estimated expected throughput on channel m' as

$$\tilde{U}_n^{m'} = \tilde{\theta}_{m'} \tilde{B}_{m'}^n \tilde{g}(k_{m'}). \tag{3.15}$$

If $\tilde{U}_n^{m'} > \tilde{U}_n^m$, then user n will imitate the channel selection of user n'.

To implement the mechanism above, each user n must have the information of its own estimated channel idle probability $\tilde{\theta}_{m'}$ and data rate $\tilde{B}_{m'}^n$ of the unchosen channel m'. Hence we add an initial channel estimation stage in the imitative spectrum access mechanism in Algorithm 5. In this stage, each user initially estimates the channel idle probability $\tilde{\theta}_m$ and data rate $\tilde{B}_m^n$ by accessing all the channels in a randomized round-robin manner. This ensures that all users do not choose the same channel at the same period. Let $\mathcal{M}_n$ (equals to the empty set $\oslash$ initially) be set of channels probed by user n and $\mathcal{M}_n^c = \mathcal{M} \setminus \mathcal{M}_n$. At beginning of each decision period, user n randomly chooses a channel $m \in \mathcal{M}_n^c$ (i.e., a channel that has not been accessed before) to access. At end of the period, user n can estimate the channel idle probability $\tilde{\theta}_m$ and data rate $\tilde{B}_m^n$ according to the MLE method introduced in Sect. 3.3.1.

Numerical results show that the proposed imitative spectrum access mechanism with user heterogeneity can still converge to an imitation equilibrium satisfying the definition in (3.2), i.e., no user can further improve its expected throughput by imitating another user. Numerical results show that the imitative spectrum access mechanism with user heterogeneity achieves up-to 500 % fairness improvement with at most 20 % performance loss, compared with the centralized optimal solution. This demonstrates that the proposed imitation-based mechanism can achieve efficient spectrum utilization and meanwhile provide good fairness across secondary users.

3.6 Simulation Results

In this section, we evaluate the proposed imitative spectrum access mechanisms by simulations. We consider a spectrum sharing network consisting $M = 5$ Rayleigh fading channels. The data rate on an idle channel m of user n is computed according to the Shannon capacity, i.e., $b_m^n = E_m \log_2(1 + \frac{\eta_n h_m^n}{n_0})$, where E_m is the bandwidth of channel m, η_n is the power adopted by user n, n_0 is the noise power, and h_m^n is the channel gain (a realization of a random variable that follows the exponential distribution with the mean $\bar{h}_m^n$). By setting different mean channel gain $\bar{h}_m^n$, we can have different mean data rates $B_m^n = E[b_m^n]$. In the following simulations, we set $\zeta_m = 10$ MHz, $n_0 = -100$ dBm, and $\eta_n = 100$ mW. We set the number of time slots in each decision period as 100. We will consider both cases with homogeneous and heterogenous users.

Algorithm 5 Imitative Spectrum Access With User Heterogeneity

---

1: **loop** for each user $n \in \mathcal{N}$ in parallel:
  ▷ *Initial Channel Estimation Stage*
2:   **while** $\mathcal{M}_n \neq \mathcal{M}$ **do**
3:     **choose** a channel m from the set $\mathcal{M}_n^c$ randomly.
4:     **sense** and **contend** to access the channel m at each time slot of the decision period.
5:     **record** the observations $S_n(t, \tau)$, $I_n(t, \tau)$ and $b_n(t, \tau)$.
6:     **estimate** the channel idle probability $\hat{\theta}_m$ and data rate $\tilde{B}_m^n$.
7:     **set** $\mathcal{M}_n = \mathcal{M}_n \cup \{m\}$.
8:   **end while**

  ▷ *Imitative Spectrum Access Stage*
9:   **for** each time period t **do**
10:     **sense** and **contend** to access the channel m at each time slot of the decision period.
11:     **record** the observations $S_n(t, \tau)$, $I_n(t, \tau)$ and $b_n(t, \tau)$.
12:     **estimate** the expected throughput $\tilde{U}_n^{a_n}(t)$.
13:     **select** another user $n' \in \mathcal{N}_n$ randomly and **enquiry** its channel grabbing probability $\tilde{g}(k_{a_{n'}})$.
14:     **estimate** the expected throughput $\tilde{U}_n^{a_{n'}}(t)$ based on (3.15).
15:     **if** $\tilde{U}_n^{a_{n'}}(t) > \tilde{U}_n^{a_n}(t)$ **then**
16:       **choose** channel $a_{n'}$ (i.e., the one chosen by user n') in the next period.
17:     **else choose** the original channel in the next period.
18:     **end if**
19:   **end for**
20: **end loop**

3.6.1 Imitative Spectrum Access with Homogeneous Users

3.6.1.1 I.i.d. Channel Environment

We first implement the imitative spectrum access mechanism with $N = 150$ homogeneous users (i.e., Algorithm 3) and the number of backoff mini-slots $\lambda_{\max} = 50$. For each user n, the mean channel data rates $\{B_m^n\}_{m=1}^M = \{15, 70, 90, 40, 100\}$ Mbps, respectively. The channel states are i.i.d. Bernoulli random variable with the mean idle probabilities $\{\theta_m\}_{m=1}^M = \{\frac{2}{3}, \frac{4}{7}, \frac{5}{9}, \frac{1}{2}, \frac{4}{5}\}$, respectively.

We consider that the information sharing graphs are represented by different cluster-based graphs as shown in Fig. 3.5. In Graph (a), clusters 1 and 3 do not communicate directly and they are connected to cluster 2. In Graph (b), all three clusters are isolated. We show the time average user's throughput in Fig. 3.6. We see that all the users achieve the same average throughput on Graph (a). This verifies the theoretic result that when information sharing graph is connected (i.e., the corresponding cluster-based graph is connected), all the users achieve the same average throughput in the imitation equilibrium. When information sharing graph is not connected (e.g., Graph (b)), we see that users in different clusters may achieve different throughputs. However, all the users in the same cluster have the same average throughput. This is also an imitation equilibrium given the constraint of their information sharing. Moreover, we see that all the channels will be utilized in the imitation

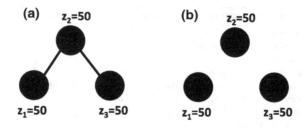

Fig. 3.5 Four types of cluster-based graphs with z_k representing the number of users in each cluster k

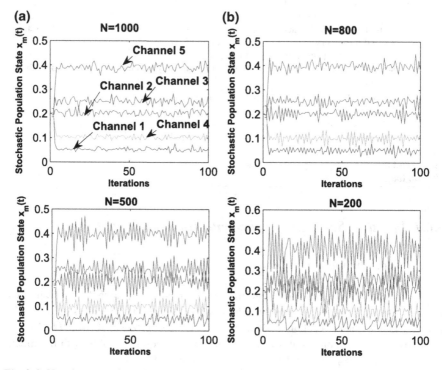

Fig. 3.6 Users' average throughputs and fractions of users on different channels on cluster-based graphs (a) and (b) in Fig. 3.5

equilibria on both Graphs (a) and (b). A channel of a higher data rate will be utilized by a larger fraction of users.

3.6.1.2 Markovian Channel Environment

For the interests of obtaining closed form solutions and deriving engineering insights, we have considered the i.i.d. channel model so far. We now evaluate the proposed mechanism in the Markovian channel environment. We denote the channel state probability vector of channel m at a time slot τ as $\boldsymbol{p}_m(\tau) \triangleq (Pr\{S_m(\tau) = 0\}, Pr\{S_m(\tau) = 1\})$, which follows a two-states Markov chain as $\boldsymbol{p}_m(\tau + 1) = \boldsymbol{p}_m(\tau)\Gamma_m, \forall \tau \geq 1$, with the transition matrix $\Gamma_m = \begin{bmatrix} 1 - p_m & p_m \\ q_m & 1 - q_m \end{bmatrix}$. In this case, we can obtain the stationary distribution that the channel m is idle with a probability of $\theta_m = \frac{p_m}{p_m + q_m}$. The study in [23] shows that the statistical properties of spectrum usage from empirical measurement data can be accurately captured and reproduced by properly setting the transition matrix. In this experiment, we choose different p_m and q_m for different channels such that the idle probabilities $\{\theta_m\}_{m=1}^{M} = \{\frac{2}{3}, \frac{4}{7}, \frac{5}{9}, \frac{1}{2}, \frac{4}{5}\}$ are the same as before. We consider that $N = 150$ users are randomly scattered across a square area of a side-length of 250 m with the information sharing graph as shown in Fig. 3.7. As mentioned in Sect. 3.4.1, this information sharing graph can also be regarded as a cluster-based graph by considering a single user as a cluster.

The results are shown in the upper part of Fig. 3.8. We observe that the imitative spectrum access mechanism still achieves the imitation equilibrium in the Markovian channel environment. The average throughput that each user achieves is the same as that in i.i.d. channel environment on the connected graph (a) in Fig. 3.6. This is because that our proposed Maximum Likelihood Estimation of the channel idle probability θ_m follows the sample average approach. By the law of large numbers, when the observation samples are sufficient, such a sample average approach can achieve an accurate estimation of the average statistics of the channel availability, even if the channel state is not an i.i.d. process.

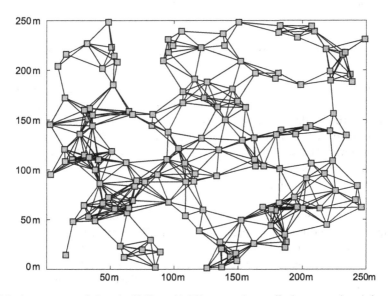

Fig. 3.7 A square area of a length of 250 m with 150 scattered users. Each user can share information with those users that are connected with it by an edge

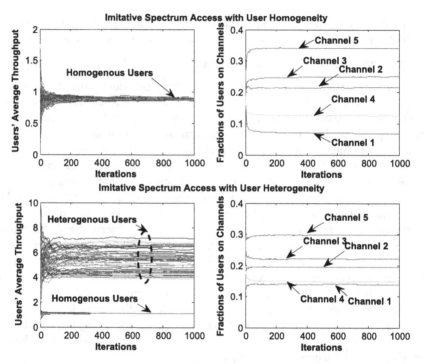

Fig. 3.8 Users' average throughputs and fractions of users on different channels on the information sharing graph in Fig. 3.7

3.6.2 Imitative Spectrum Access with Heterogeneous Users

We then implement the imitation spectrum access mechanism with heterogeneous users (i.e., Algorithm 5) on the same information sharing graph in Fig. 3.7. The mean data rates of 100 randomly chosen users out of these 150 users are homogeneous with the same data rates as before (i.e., $\{B^n_m\}^M_{m=1} = \{15, 70, 90, 40, 100\}$ Mbps). For the remaining 50 users, we set that the users' data rates are heterogenous with the mean data rate of user n on channel m as $B^n_m = 100 + R$ where R is a random value drawn from the uniform distribution over $(0, 100)$. The results are shown in the bottom part of Fig. 3.8. We see that the mechanism converges to the equilibrium wherein homogeneous users achieve the same expected throughput and heterogenous users may achieve different expected throughputs. Moreover, we observe that the mechanism converges to a stable user distribution on channels. This implies that no user can further improve its expected throughput by imitating another user. That is, the equilibrium is an imitation equilibrium satisfying the definition in (3.13).

3.6.3 Performance Comparison

We now compare the proposed imitative spectrum access mechanism with the imitation-based spectrum access mechanism in [4]. Notice that the mechanism in [4] requires the global network information including the channel characteristics and other users' channel selections to compute user's throughput. When users are homogenous (i.e., different users achieve the same data rate on the same channel), both mechanisms can converge to the imitation equilibrium and hence achieve the same performance. Due to space limit, please refer to [5] for the details of performance comparison in the homogenous user case.

We now focus on performance comparisons for the more general and practical case that users are heterogenous. As the benchmark, we also implement the centralized optimal solution that maximizes the system-wide throughput (i.e., $\max \sum_{n=1}^{N} U_n$) and the decentralized spectrum access solution by Q-learning mechanism proposed in [24]. Similarly to the setting in Sect. 3.6.2, we consider $N = 100, 150, \ldots, 300$ randomly scattered users, respectively. The mean data rate of user n on channel m is $B_m^n = R$, where R is a random value drawn from the uniform distribution over $(0, 200)$. For each fixed user number N, we average over 50 runs.

We first show the system-wide throughput achieved by different mechanisms in Fig. 3.9. We see that the system-wide throughput of all the solutions decreases as the number of users increases. As the user population increases, the contention among users becomes more severe, which leads to more spectrum access collisions. We observe that the proposed imitative spectrum access mechanism with user heterogeneity achieves up-to 32 % performance improvement over the imitation-based spectrum access mechanism in [4]. This is because that the mechanism in [4] carries out imitation based on other user's throughput information directly, which ignores the fact that users are heterogeneous. While our mechanism takes user heterogeneity into account and carries out imitation based on the channel contention level. Compared with Q-learning mechanism, the imitative spectrum access mechanism can achieve better performance, with a performance gain of around 5 %. Moreover, the performance loss of the imitative spectrum access mechanism with respect to the centralized optimal solution is at most 20 % in all cases. This demonstrates the efficiency of the imitative spectrum access mechanism with user heterogeneity.

We then compare the fairness achieved by different mechanisms in Fig. 3.10. We adopt the widely-used Jain's fairness index [22] to measure the fairness. A larger index J represents a more fair channel allocation, with the best case $J = 1$. Figure 3.10 shows that the centralized optimal solution is poor in terms of fairness (with the highest index value $J = 0.2$). This reason is that the centralized optimal solution would allocate the best channels to a small fraction of users only (to avoid congestion) and most users will share those channels of low data rates. Our imitative spectrum access is much more fair and achieves up to 530 and 300 % fairness improvement over the centralized optimization and Q-learning, respectively. This demonstrates that the proposed imitation-based mechanism can provide good fairness across users.

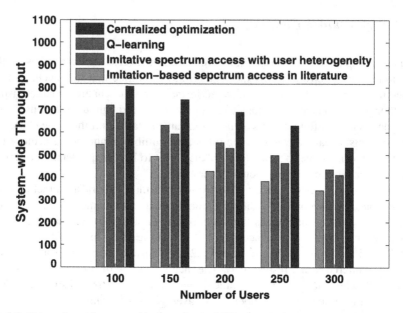

Fig. 3.9 Comparison of system-wide throughput of different solutions

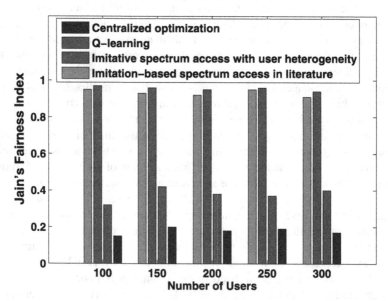

Fig. 3.10 Comparison of fairness of the solutions of different solutions

3.7 Summary

In this chapter, we design a distributed spectrum access mechanism with incomplete network information based on social imitations. We show that the imitative spectrum access mechanism can converge to an imitation equilibrium on different information sharing graphs. When the information sharing graph is connected and users are homogeneous, the imitation equilibrium corresponds to a fair channel allocation such that all the users achieve the same throughput. We also extend the imitative spectrum access mechanism to the case that users are heterogeneous. Numerical results demonstrate that the proposed imitation-based mechanism can achieve efficient spectrum utilization and meanwhile provide good fairness across secondary users.

References

1. W. Wyrwicka, *Imitation in Human and Animal Behavior* (Transaction Publishers, New Brunswick, 1996)
2. K. Schlag, Why imitate, and if so, how? a boundedly rational approach to multi-armed bandits. J. Econ. Theory **78**, 130–156 (1998)
3. M. Lopes, F.S. Melo, L. Montesano, Affordance-based imitation learning in robots, in *IEEE/RSJ International Conference on Intelligent Robots and Systems* (2007)
4. S. Iellamo, L. Chen, M. Coupechoux, Let cognitive radios imitate: imitation-based spectrum access for cognitive radio networks, Technical Report, arXiv:1101.6016 (2011)
5. X. Chen, J. Huang, Imitation-based social spectrum sharing. IEEE Trans. Mob. Comput. (2014). http://arxiv.org/pdf/1405.2822v1.pdf
6. P. Gupta, P. Kumar, The capacity of wireless networks. IEEE Trans. Inf. Theory **46**(2), 388–404 (2000)
7. A.B. Flores, R.E. Guerra, E.W. Knightly, P. Ecclesine, S. Pandey, IEEE 802.11 af: a standard for TV white space spectrum sharing. IEEE Commun. Mag. **51**(10), 92–100 (2013)
8. R.J. Drost, R.D. Hopkins, R. Ho, I.E. Sutherland, Proximity communication. IEEE J. Solid-State Circuits **39**(9), 1529–1535 (2004)
9. L. Anderegg, S. Eidenbenz, Ad hoc-VCG: a truthful and cost-efficient routing protocol for mobile ad hoc networks with selfish agents, in *9th Annual International Conference on Mobile Computing And Networking* (2003)
10. P. Michiardi, R. Molva, Core: a collaborative reputation mechanism to enforce node cooperation in mobile ad hoc networks, in *Advanced Communications and Multimedia Security* (2002)
11. L.A. DaSilva, I. Guerreiro, Sequence-based rendezvous for dynamic spectrum access, in *3rd IEEE Symposium on New Frontiers in Dynamic Spectrum Access Networks* (2008)
12. K. Bian, J.-M. Park, R. Chen, Control channel establishment in cognitive radio networks using channel hopping. IEEE J. Sel. Areas Commun. **29**(4), 689–703 (2011)
13. L. Lazos, S. Liu, M. Krunz, Spectrum opportunity-based control channel assignment in cognitive radio networks, in *6th Annual IEEE Communications Society Conference on Sensor, Mesh and Ad Hoc Communications and Networks* (2009)
14. B. Lo, A survey of common control channel design in cognitive radio networks. Phys. Commun. **4**, 40–62 (2011)
15. R. Zhang, Y. Zhang, J. Sun, G. Yan, Fine-grained private matching for proximity-based mobile social networking, in *IEEE INFOCOM* (2012)
16. M. Von Arb, M. Bader, M. Kuhn, R. Wattenhofer, Veneta: serverless friend-of-friend detection in mobile social networking, in *IEEE International Conference on Wireless and Mobile Computing, Networking and Communications* (2008)

17. B. Cohen, Incentives build robustness in bittorrent. Workshop Econ. Peer-to-Peer Syst. **6**, 68–72 (2003)
18. T. Ferguson, *A Course in Large Sample Theory* (Chapman and Hall, London, 1996)
19. C. Alós-Ferrer, S. Weidenholzer, Contagion and efficiency. J. Econ. Theory **143**(1), 251–274 (2008)
20. J. Scott, Social network analysis. Sociology **22**(1), 109 (1988)
21. M. Benam, J. Weibull, Deterministic approximation of stochastic evolution in games. Econometrica **71**, 873–903 (2003)
22. R. Jain, D. Chiu, W. Hawe, A quantitative measure of fairness and discrimination for resource allocation in shared computer systems, DEC Research Report TR-301 (1984)
23. M. Lopez-Benitez, F. Casadevall, Empirical time-dimension model of spectrum use based on a discrete-time markov chain with deterministic and stochastic duty cycle models. IEEE Trans. Veh. Technol. **60**(6), 2519–2533 (2011)
24. H. Li, Multi-agent Q-learning for competitive spectrum access in cognitive radio systems, in *Fifth IEEE Workshop on Networking Technologies for Software Defined Radio (SDR) Networks* (2010)
25. A. Anandkumar, N. Michael, A. Tang, Opportunistic spectrum access with multiple users: learning under competition, in *IEEE INFOCOM* (2010)
26. T. Rappaport, *Wireless Communications: Principles and Practice*, vol. 2 (Prentice Hall PTR, New Jersey, 1996)

Chapter 4
Evolutionarily Stable Spectrum Access Mechanism

4.1 Introduction

In this chapter, motivated by the evolution rule observed in many social animal and human interactions, we propose a new framework of distributed spectrum access with and without complete network information (i.e., channel statistics and user selections). A key feature of this framework is that we design the spectrum access mechanisms based on *bounded rationality* of secondary users, i.e., each user tries to improve its strategy adaptively over time, which requires much less computation power than the full rationality case, and thus may better match the reality of wireless communications. This framework has a sharp difference with the approaches of using non-cooperative game for distributed spectrum access in which secondary users are *fully rational* and thus often adopt their channel selections based on best responses, i.e., the best choices they can compute by having the complete network information. To have full rationality, a user needs to have a high computational power to collect and analyze the network information in order to predict other users' behaviors. This is often not feasible due to the limitations of today's wireless devices.

We first propose an evolutionary game approach for distributed spectrum access with the complete network information, where each secondary user takes a *comparison* strategy (i.e., comparing its payoff with the system average payoff) to evolve its spectrum access decision over time. We then propose a learning mechanism for distributed spectrum access with incomplete information, which does not require any prior knowledge of channel statistics or information exchange among users. In this case, each secondary user estimates its expected throughput locally, and *learns* to adjust its channel selection strategy adaptively.

The main results and contributions of this chapter are as follows:

- *Evolutionary spectrum access mechanism*: we formulate the distributed spectrum access over multiple heterogeneous time-varying licensed channels as an evolutionary spectrum access game, and study the evolutionary dynamics of spectrum access.

© The Author(s) 2015
X. Chen and J. Huang, *Social Cognitive Radio Networks*,
SpringerBriefs in Electrical and Computer Engineering,
DOI 10.1007/978-3-319-15215-8_4

- *Evolutionary dynamics and stability*: we show that the evolutionary spectrum access mechanism converges to the evolutionary equilibrium, and prove that it is globally evolutionarily stable.
- *Learning mechanism with incomplete information*: we further propose a learning mechanism without the knowledge of channel statistics and user information exchange. We show that the learning mechanism converges to the same evolutionary equilibrium on the time average.
- *Superior performance*: we show that the proposed mechanisms can achieve up to 35 % performance improvement over the distributed reinforcement learning mechanism in literature, and are robust to the perturbations of users' channel selections.

The rest of the chapter is organized as follows. We introduce the system model in Sect. 4.2. After briefly reviewing the evolutionary game theory in Sect. 4.3, we present the evolutionary spectrum access mechanism with complete information in Sect. 4.4. Then we introduce the learning mechanism in Sect. 4.5. We illustrate the performance of the proposed mechanisms through numerical results in Sect. 4.6 and finally conclude in Sect. 4.7. *Due to space limitations, the details for several proofs are provided in* [1].

4.2 System Model

We consider a cognitive radio network with a set $\mathcal{M} = \{1, 2, \ldots, M\}$ of independent and *stochastically heterogeneous* licensed channels. A set $\mathcal{N} = \{1, 2, \ldots, N\}$ of secondary users try to opportunistically access these channels, when the channels are not occupied by primary (licensed) transmissions. The system model has a slotted transmission structure as in Fig. 4.1 and is described as follows.

- *Channel State*: the channel state for a channel m at time slot t is

$$S_m(t) = \begin{cases} 0, & \text{if channel } m \text{ is occupied by} \\ & \text{primary transmissions,} \\ 1, & \text{if channel } m \text{ is idle.} \end{cases}$$

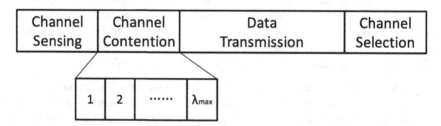

Fig. 4.1 Multiple stages in a single time slot

- *Channel State Changing*: for a channel m, we assume that the channel state is an i.i.d. Bernoulli random variable, with an idle probability $\theta_m \in (0, 1)$ and a busy probability $1 - \theta_m$. This model can be a good approximation of the reality if the time slots for secondary transmissions are sufficiently long or the primary transmissions are highly bursty [3]. Numerical results show that the proposed mechanisms also work well in the Markovian channel environment.
- *Heterogeneous Channel Throughput*: if a channel m is idle, the achievable data rate $b_m(t)$ by a secondary user in each time slot t evolves according to an i.i.d. random process with a mean B_m, due to the local environmental effects such fading. For example, in a frequency-selective Rayleigh fading channel environment we can compute the channel data rate according to the Shannon capacity with the channel gain at a time slot being a realization of a random variable that follows the exponential distribution [4].
- *Time Slot Structure*: each secondary user n executes the following stages synchronously during each time slot:

  - *Channel Sensing*: sense one of the channels based on the channel selection decision generated at the end of previous time slot. Access the channel if it is idle.
  - *Channel Contention*: use a backoff mechanism to resolve collisions when multiple secondary users access the same idle channel. The contention stage of a time slot is divided into λ_{max} mini-slots[1] (see Fig. 4.1), and user n executes the following two steps. *First*, count down according to a randomly and uniformly chosen integral backoff time (number of mini-slots) λ_n between 1 and λ_{max}. *Second*, once the timer expires, transmit RTS/CTS messages if the channel is clear (i.e., no ongoing transmission). Note that if multiple users choose the same backoff value λ_n, a collision will occur with RTS/CTS transmissions and no users can successfully grab the channel.
  - *Data Transmission*: transmit data packets if the RTS/CTS message exchanges go through and the user successfully grabs the channel.
  - *Channel Selection*: in the complete information case, users broadcast the chosen channel IDs to other users through a common control channel,[2] and then make the channel selection decisions based on the evolutionary spectrum access mechanism (details in Sect. 4.4). With incomplete information, users update the channel estimations based on the current access results, and make the channel selection decisions according to the distributed learning mechanism (details in Sect. 4.5).

---

[1] For the ease of exposition, we assume that the contention backoff size λ_{max} is fixed. This corresponds to an equilibrium model for the case that the backoff size λ_{max} can be dynamically tuned according to the 802.11 distributed coordination function [5]. Also, we can enhance the performance of the backoff mechanism by determining optimal fixed contention backoff size according to the method in [6].

[2] Please refer to [7] for the details on how to set up and maintain a reliable common control channel in cognitive radio networks.

Suppose that k_m users choose an idle channel m to access. Then the probability that a user n (out of the k_m users) grabs the channel m is

$$g(k_m) = Pr\{\lambda_n < \min_{i \neq n}\{\lambda_i\}\}$$

$$= \sum_{\lambda=1}^{\lambda_{\max}} Pr\{\lambda_n = \lambda\} Pr\{\lambda < \min_{i \neq n}\{\lambda_i\}|\lambda_n = \lambda\}$$

$$= \sum_{\lambda=1}^{\lambda_{\max}} \frac{1}{\lambda_{\max}} \left(\frac{\lambda_{\max} - \lambda}{\lambda_{\max}}\right)^{k_m - 1},$$

which is a decreasing function of the number of total contending users k_m. Then the expected throughput of a secondary user n choosing a channel m is given as

$$U_n = \theta_m B_m g(k_m). \tag{4.1}$$

For the ease of exposition, we will focus on the analysis of the proposed spectrum access mechanisms in the many-users regime. Numerical results show that our algorithms also apply when the number of users is small (see Sect. 4.6.3 for the details). Since our analysis is from secondary users' perspective, we will use terms "secondary user" and "user" interchangeably.

4.3 Overview of Evolutionary Game Theory

For the sake of completeness, we will briefly describe the background of evolutionary game theory. Detailed introduction can be found in [8]. Evolutionary game theory was first used in biology to study the change of animal populations, and then later applied in economics to model human behaviors. It is most useful to understand how a large population of users converge to Nash equilibria in a dynamic system [8]. A player in an evolutionary game has bounded rationality, i.e., limited computational capability and knowledge, and improves its decisions as it learns about the environment over time [8].

4.3.1 Replicator Dynamics

As mentioned, the evolutionary game theory was first proposed in Biology to study the interactive behaviors among a population of animals [8]. The game consists of animals (players) using different strategies. The strategy of an animal is inherited by its offsprings. Animals with higher fitness will leave more offsprings, so in the next generation the composition of the population will change. Such a reproduction process can be modeled by a set of differential equations called replicator dynamics.

Formally, a player in the population chooses a strategy i from a finite set of strategies $\mathscr{I} = \{1, \ldots, I\}$. The population state $\mathbf{x}(t) = (x_1(t), \ldots, x_I(t))$ describes the dynamics of the reproduction process, with $x_i(t)$ denoting the proportion of players in the population adopting the strategy i at time t. The replicator dynamics are then given by

$$\dot{x}_i(t) = \beta \left(R(i, \mathbf{x}(t)) - R(\mathbf{x}(t)) \right), \quad \forall i \in \mathscr{I}, \tag{4.2}$$

where $R(i, \mathbf{x}(t))$ is the payoff of the players choosing strategy i, $R(\mathbf{x}(t))$ is the average payoff of the population, and $\beta > 0$ is the rate of strategy adaptation. It means that the strategy that works better than the average will be promoted.

4.3.2 Evolutionarily Stable Strategy

The evolutionarily stable strategy (ESS) is a key concept to describe the evolutionary equilibrium. For simplicity, we will introduce the ESS definition (the strict Nash equilibrium in Definition 4.2, respectively) in the context of a symmetric game where all users adopt the same strategy i at the ESS (strict Nash equilibrium, respectively). The definition can be (and will be) extended to the case of asymmetric game [8], where we view the population's collective behavior as a mixed strategy i at the ESS (strict Nash equilibrium, respectively).

An ESS ensures the stability such that the population is robust to perturbations by a small fraction of players. Suppose that a small share ε of players in the population deviate to choose a mutant strategy j, while all other players stick to the incumbent strategy i. We denote the population state of the game as $\mathbf{x}_{(1-\varepsilon)i+\varepsilon j} = \left(x_i = 1 - \varepsilon, x_j = \varepsilon, x_l = 0, \forall l \neq i, j \right)$, where x_a denotes the fraction of users choosing strategy a, and the corresponding payoff of choosing strategy a as $R(a, \mathbf{x}_{(1-\varepsilon)i+\varepsilon j})$.

Definition 4.1 ([8]) A strategy i is an **evolutionarily stable strategy** if for every strategy $j \neq i$, there exists an $\bar{\varepsilon} \in (0, 1)$ such that $R(i, \mathbf{x}_{\varepsilon j+(1-\varepsilon)i}) > R(j, \mathbf{x}_{\varepsilon j+(1-\varepsilon)i})$ for any $j \neq i$ and $\varepsilon \in (0, \bar{\varepsilon})$.

Definition 4.1 means that the mutant strategy j cannot invade the population when the perturbation is small enough, if the incumbent strategy i is an ESS. It is shown in [8] that any strict Nash equilibrium in noncooperative games is also an ESS.

Definition 4.2 ([8]) A strategy i is a **strict Nash equilibrium** if for every strategy $j \neq i$, it satisfies that $R(i, i, \ldots, i) > R(j, i, \ldots, i)$, where $R(a, i, \ldots, i)$ denotes the payoff of choosing strategy $a \in \{i, j\}$ given other players adhering to the strategy i.

To understand that a strict Nash is an ESS, we can set $\varepsilon \to 0$ in Definition 4.1, which leads to $R(i, \mathbf{x}_i) > R(j, \mathbf{x}_i), \forall j \neq i$, i.e., given that almost all other players play the incumbent strategy i, choosing any mutant strategy $j \neq i$ will lead to a loss in payoff.

Several recent results applied the evolutionary game theory to study various networking problems. Niyato and Hossain [3] investigated the evolutionary dynamics of heterogeneous network selections. Zhang et al. [9] designed incentive schemes for resource-sharing in P2P networks based on the evolutionary game theory. Wang et al. [10] proposed the evolutionary game approach for collaborative spectrum sensing mechanism design in cognitive radio networks. According to Definition 4.1, the ESS obtained in [3, 9, 10] is locally evolutionarily stable (i.e., the mutation ε is small enough). Here we apply the evolutionary game theory to design spectrum access mechanism, which can achieve global evolutionary stability (i.e., the mutation ε can be arbitrarily large).

4.4 Evolutionary Spectrum Access

We now apply the evolutionary game theory to design an efficient and stable spectrum access mechanism with complete network information. We will show that the spectrum access equilibrium is an ESS, which guarantees that the spectrum access mechanism is robust to random perturbations of users' channel selections.

4.4.1 Evolutionary Game Formulation

The evolutionary spectrum access game is formulated as follows:

- Players: the set of users $\mathcal{N} = \{1, 2, \ldots, N\}$.
- Strategies: each user can access one of the set of channels $\mathcal{M} = \{1, 2, \ldots, M\}$.
- Population state: the user distribution over M channels at time t, $\boldsymbol{x}(t) = (x_m(t), \forall m \in \mathcal{M})$, where $x_m(t)$ is the proportion of users selecting channel m at time t. We have $\sum_{m \in \mathcal{M}} x_m(t) = 1$ for all t.
- Payoff: a user n's expected throughput $U_n(a_n, \boldsymbol{x}(t))$ when choosing channel $a_n \in \mathcal{M}$, given that the population state is $\boldsymbol{x}(t)$. Since each user has the information of channel statistics, from (4.1), we have

$$U_n(a_n, \boldsymbol{x}(t)) = \theta_{a_n} B_{a_n} g(N x_{a_n}(t)). \tag{4.3}$$

We also denote the system arithmetic average payoff under population state $\boldsymbol{x}(t)$ as

$$U(\boldsymbol{x}(t)) = \frac{1}{M} \sum_{m=1}^{M} \theta_m B_m g(N x_m(t)). \tag{4.4}$$

Algorithm 6 Evolutionary Spectrum Access Mechanism

1: **initialization:**
2:    **set** the global strategy adaptation factor $\alpha \in (0, 1]$.
3:    **select** a random channel for each user.
4: **end initialization**

5: **loop** for each time slot t and each user $n \in \mathcal{N}$ in parallel:
6:    **sense** and **contend** for the chosen channel and transmit data packets if successfully grabbing the channel.
7:    **broadcast** the chosen channel ID to other users through a common control channel.
8:    **receive** the information of other users' channel selection and calculate the population state $x(t)$.
9:    **compute** the expected payoff $U_n(a_n, x(t))$ and the system average payoff $U(x(t))$ according to (4.3) and (4.4), respectively.
10:    **if** $U_n(a_n, x(t)) < U(x(t))$ **then**
11:       **generate** a random value δ according to a uniform distribution on $(0, 1)$.
12:       **if** $\delta < \frac{\alpha}{x_{a_n}(t)} \left(1 - \frac{U_n(a_n, x(t))}{U(x(t))}\right)$ **then**
13:          **select** a new channel m with probability

$$p_m = \frac{\max\{\theta_m B_m g(N x_m(t)) - U(x(t)), 0\}}{\sum_{m'=1}^{M} \max\{\theta_{m'} B_{m'} g(N x_{m'}(t)) - U(x(t)), 0\}}.$$

14:       **else select** the original channel.
15:       **end if**
16:    **end if**
17: **end loop**

4.4.2 Evolutionary Dynamics

Based on the evolutionary game formulation above, we propose an evolutionary spectrum access mechanism in Algorithm 6 by reversing-engineering the replicator dynamics. The idea is to let those users who have payoffs lower than the system average payoff $U(x(t)$ to select a better channel, with a probability proportional to the (normalized) channel's "net fitness" $\theta_m B_m g(N x_m(t)) - U(x(t))$. We show that the dynamics of channel selections in the mechanism can be described with the evolutionary dynamics in (4.5).

Theorem 4.1 *For the evolutionary spectrum access mechanism in Algorithm 6, the evolutionary dynamics are given as*

$$\dot{x}_m(t) = \alpha \left(\frac{U_n(m, x(t))}{U(x(t))} - 1\right), \quad \forall m \in \mathcal{M}, \tag{4.5}$$

where the derivative is with respect to time t.

4.4.3 Evolutionary Equilibrium in Asymptotic Case $\lambda_{max} = \infty$

We next investigate the equilibrium of the evolutionary spectrum access mechanism. To obtain useful insights, we first focus on the asymptotic case where the number of backoff mini-slots λ_{max} goes to ∞, such that

$$
\begin{aligned}
g(k) &= \lim_{\lambda_{max} \to \infty} \sum_{\lambda=1}^{\lambda_{max}} \frac{1}{\lambda_{max}} \left(\frac{\lambda_{max} - \lambda}{\lambda_{max}} \right)^{k-1} \\
&= \lim_{\frac{1}{\lambda_{max}} \to 0} \sum_{\lambda=0}^{\lambda_{max}-1} \left(\frac{\lambda}{\lambda_{max}} \right)^{k-1} \frac{1}{\lambda_{max}} \\
&= \int_0^1 z^{k-1} dz = \frac{1}{k}.
\end{aligned}
\tag{4.6}
$$

This is a good approximation when the number of mini-slots λ_{max} for backoff is much larger than the number of users N and collisions rarely occur. In this case, $U_n(a_n, \boldsymbol{x}(t)) = \frac{\theta_{a_n} B_{a_n}}{N x_m(t)}$ and $U(\boldsymbol{x}(t)) = \frac{\sum_{i=1}^{M} \theta_i B_i}{N}$. According to Theorem 4.1, the evolutionary dynamics in (4.5) become

$$
\dot{x}_m(t) = \alpha \left(\frac{\frac{\theta_m B_m}{x_m(t)}}{\frac{1}{M} \sum_{i=1}^{M} \frac{\theta_i B_i}{x_i(t)}} - 1 \right).
\tag{4.7}
$$

From (4.7), we have

Theorem 4.2 *The evolutionary spectrum access mechanism in asymptotic case* $\lambda_{max} = \infty$ *globally converges to the evolutionary equilibrium* $\boldsymbol{x}^* = \left(x_m^* = \frac{\theta_m B_m}{\sum_{i=1}^{M} \theta_i B_i}, \forall m \in \mathcal{M} \right)$.

Theorem 4.2 implies that

Lemma 4.1 *The evolutionary spectrum access mechanism converges to the equilibrium* $\boldsymbol{x}^*$ *such that users on different channels achieve the same expected throughput, i.e.,*

$$
U_n(m, \boldsymbol{x}^*) = U_n(m', \boldsymbol{x}^*), \quad \forall m, m' \in \mathcal{M}.
\tag{4.8}
$$

We next show that for the general case $\lambda_{max} < \infty$, the evolutionary dynamics also globally converges to the ESS equilibrium as given in (4.8).

4.4.4 Evolutionary Equilibrium in General Case $\lambda_{max} < \infty$

For the general case λ_{max}, since the channel grabbing probability $g(k)$ does not have the close-form expression, it is hence difficult to obtain the equilibrium solution of differential equations in (4.5). However, it is easy to verify that the equilibrium x^* in (4.8) is also a stationary point such that the evolutionary dynamics (4.5) in the general case $\lambda_{max} < \infty$ satisfy $\dot{x}_m(t) = 0$. Thus, at the equilibrium x^*, users on different channels achieve the same expected throughput.

We now study the evolutionary stability of the equilibrium. In general, the equilibrium of the replicator dynamics may not be an ESS [8]. For our model, we can prove the following.

Theorem 4.3 *For the evolutionary spectrum access mechanism, the evolutionary equilibrium x^* in (4.8) is an ESS.*

Actually we can obtain a stronger result than Theorem 4.3. Typically, an ESS is only locally asymptotically stable (i.e., stable within a limited region around the ESS) [8]. For our case, we show that the evolutionary equilibrium x^* is globally asymptotically stable (i.e., stable in the entire feasible region of a population state x, $\{x = (x_m, m \in \mathcal{M}) | \sum_{m=1}^{M} x_m = 1 \text{ and } x_m \geq 0, \forall m \in \mathcal{M}\}$).

To proceed, we first define the following function

$$L(x) = \sum_{m=1}^{M} \int_{-\infty}^{x_m} \theta_m B_m g(Nz) dz. \tag{4.9}$$

Since $g(\cdot)$ is a decreasing function, it is easy to check that the Hessian matrix of $L(x)$ is negative definite. It follows that $L(x)$ is strictly concave and hence has a unique global maximum L^*. By the first order condition, we obtain the optimal solution x^*, which is the same as the evolutionary equilibrium x^* in (4.8). Then by showing that $V(x(t)) = L^* - L(x(t))$ is a strict Lyapunov function, we have

Theorem 4.4 *For the evolutionary spectrum access mechanism, the evolutionary equilibrium x^* in (4.8) is globally asymptotically stable.*

Since the ESS is globally asymptotically stable, the evolutionary spectrum access mechanism is robust to any degree of (not necessarily small) random perturbations of channel selections.

4.5 Learning Mechanism for Distributed Spectrum Access

For the evolutionary spectrum access mechanism in Sect. 4.4, we assume that each user has the perfect knowledge of channel statistics and the population state by information exchange on a common control channel. Such mechanism leads to significant communication overhead and energy consumption, and may even be impossible in some systems. We thus propose a learning mechanism for distributed spectrum access

---

Algorithm 7 Learning Mechanism For Distributed Spectrum Access

---
1: **initialization:**
2:     set the global memory weight $\gamma \in (0, 1)$ and the set of accessed channels $\mathcal{M}_n = \varnothing$ for
     each user n.
3: **end initialization**

4: **loop** for each user $n \in \mathcal{N}$ in parallel:

   ▷ *Initial Channel Estimation Stage*
5:     **while** $\mathcal{M}_n \neq \mathcal{M}$ **do**
6:         **choose** a channel m from the set $\mathcal{M}_n^c$ randomly.
7:         **sense** and **contend** to access the channel m at each time slot of the decision period.
8:         **estimate** the expected throughput $\tilde{U}_{m,n}(0)$ by (4.10).
9:         **set** $\mathcal{M}_n = \mathcal{M}_n \cup \{m\}$.
10:     **end while**

   ▷ *Access Strategy Learning Stage*
11:     **for** for each time period T **do**
12:         **choose** a channel m to access according to the mixed strategy $f_n(T)$ in (4.11).
13:         **sense** and **contend** to access the channel m at each time slot of the decision period.
14:         **estimate** the qualities of the chosen channel m and the unchosen channels $m' \neq m$ by
             (4.13) and (4.12), respectively.
15:     **end for**
16: **end loop**

---

with incomplete information. The challenge is how to achieve the evolutionarilybreak
stable state based on user's local observations only.

4.5.1 Learning Mechanism for Distributed Spectrum Access

The proposed learning process is shown in Algorithm 7 and has two sequential stages:
initial channel estimation (line 5 to 10) and *access strategy learning* (line 11 to 15).
Each stage is defined over a sequence of decision periods $T = 1, 2, \ldots$, where each
decision period consists of t_{max} time slots (see Fig. 4.2 as an illustration).

The key idea of distributed learning here is to adapt each user's spectrum access
decision based on its accumulated experiences. In the first stage, each user initially

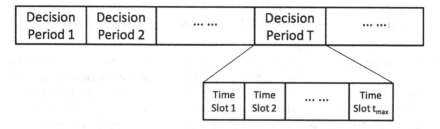

Fig. 4.2 Learning time structure

estimates the expected throughput by accessing all the channels in a randomized round-robin manner. This ensures that all users do not choose the same channel at the same period. Let $\mathcal{M}_n$ (equals to $\varnothing$ initially) be the set of channels accessed by user n and $\mathcal{M}_n^c = \mathcal{M} \setminus \mathcal{M}_n$. At beginning of each decision period, user n randomly chooses a channel $m \in \mathcal{M}_n^c$ (i.e., a channel that has not been accessed before) to access. At end of the period, user n can estimate the expected throughput by sample averaging as

$$Z_{m,n}(0) = (1 - \gamma)\frac{\sum_{t=1}^{t_{\max}} b_m(t) I_{\{a_n(t,T)=m\}}}{t_{\max}}, \tag{4.10}$$

where $0 < \gamma < 1$ is called the memory weight and $I_{\{a_n(t,T)=m\}}$ is an indicator function and equals 1 if the channel m is idle at time slot t and the user n chooses and successfully grabs the channel m. Motivation of multiplying $(1 - \gamma)$ in (4.10) is to scale down the impact of the noisy instantaneous estimation on the learning. Note that there are $t_{\max}$ time slots within each decision period, and thus the user will be able to have a fairly good estimation of the expected throughput if $t_{\max}$ is reasonably large. Then user n updates the set of accessed channels as $\mathcal{M}_n = \mathcal{M}_n \cup \{m\}$. When all the channels are accessed, i.e., $\mathcal{M}_n = \mathcal{M}$, the stage of initial channel estimation ends. Thus, the total time slots for the first stage is $M t_{\max}$.

In the second stage, at each period $T \geq 1$, each user $n \in \mathcal{N}$ selects a channel m to access according to a mixed strategy $f_n(T) = (f_{1,n}(T), \ldots, f_{M,n}(T))$, where $f_{m,n}(T)$ is the probability of user n choosing channel m and is computed as

$$f_{m,n}(T) = \frac{\sum_{\tau=0}^{T-1} \gamma^{T-\tau-1} Z_{m,n}(\tau)}{\sum_{i=1}^{M} \sum_{\tau=0}^{T-1} \gamma^{T-\tau-1} Z_{i,n}(\tau)}, \quad \forall m \in \mathcal{M}. \tag{4.11}$$

Here $Z_{m,n}(\tau)$ is user n's estimation of the quality of channel m at period τ (see (4.12) and (4.13) later). The update in (4.11) means that each user adjusts its mixed strategy according to its weighted average estimations of all channels' qualities.

Suppose that user n chooses channel m to access at period τ. For the unchosen channels $m' \neq m$ at this period, user n can empirically estimate the quality of this channel according to its past memories as

$$Z_{m',n}(\tau) = (1 - \gamma) \sum_{\tau'=0}^{\tau-1} \gamma^{\tau-\tau'-1} Z_{m',n}(\tau'). \tag{4.12}$$

For the chosen channel m, user n will update the estimation of this channel m by combining the empirical estimation with the real-time throughput measurement in this period, i.e.,

$$Z_{m,n}(\tau) = (1 - \gamma)\left(\sum_{\tau'=0}^{\tau-1} \gamma^{\tau-\tau'-1} Z_{m,n}(\tau') + \frac{\sum_{t=1}^{t_{\max}} b_m(t) I_{\{a_n(t,\tau)=m\}}}{t_{\max}}\right). \tag{4.13}$$

4.5.2 Convergence of Learning Mechanism

We now study the convergence of the learning mechanism. Since each user only utilizes its local estimation to adjust its mixed channel access strategy, the exact ESS is difficult to achieve due to the random estimation noise. We will show that the learning mechanism can converge to the ESS on time average.

According to the theory of stochastic approximation [11], the limiting behaviors of the learning mechanism with the random estimation noise can be well approximated by the corresponding mean dynamics. We thus study the mean dynamics of the learning mechanism. To proceed, we define the mapping from the mixed channel access strategies $f(T) = (f_1(T), \ldots, f_N(T))$ to the mean throughput of user n choosing channel m as $Q_{m,n}(f(T)) \triangleq E[U_n(m, x(T))|f(T)]$. Here the expectation $E[\cdot]$ is taken with respective to the mixed strategies $f(T)$ of all users. We show that

Theorem 4.5 *As the memory weight* $\gamma \to 1$, *the mean dynamics of the learning mechanism for distributed spectrum access are given as* $(\forall m \in \mathcal{M}, n \in \mathcal{N})$

$$\dot{f}_{m,n}(T) = f_{m,n}(T)\left(Q_{m,n}(f(T)) - \sum_{i=1}^{M} f_{i,n}(T)Q_{i,n}(f(T)) \right), \qquad (4.14)$$

where the derivative is with respect to period T.

Interestingly, similarly with the evolutionary dynamics in (4.5), the learning dynamics in (4.14) imply that if a channel offers a higher throughput for a user than the user's average throughput over all channels, then the user will exploit that channel more often in the future learning. However, the evolutionary dynamics in (4.5) are based on the population level with complete network information, while the learning dynamics in (4.14) are derived from the individual local estimations. We show in Theorem 4.6 that the mean dynamics of learning mechanism converge to the ESS in (4.8), i.e., $Q_{m,n}(f^*) = Q_{m',n}(f^*)$.

Theorem 4.6 *As the memory weight* $\gamma \to 1$, *the mean dynamics of the learning mechanism for distributed spectrum access asymptotically converge to a limiting point* f^* *such that*

$$Q_{m,n}(f^*) = Q_{m',n}(f^*), \quad \forall m, m' \in \mathcal{M}, \forall n \in \mathcal{N}. \qquad (4.15)$$

Since $Q_{m,n}(f^*) = E[U_n(m, x(T))|f^*]$ and the mean dynamics converge to the equilibrium f^* satisfying (4.15) (i.e., $E[U_n(m, x(T))|f^*] = E[U_n(m', x(T))|f^*]$), the learning mechanism thus converges to the ESS (4.8) (achieved by the evolutionary spectrum access mechanism) on the time average. Note that both the evolutionary spectrum access mechanism in Algorithm 6 and learning mechanism in Algorithm 7 involve basic arithmetic operations and random number generation over M channels, and hence have a linear computational complexity of $\mathcal{O}(M)$ for each iteration. However, due to the incomplete information, the learning mechanism typically takes a longer convergence time in order to get a good estimation of the environment.

4.6 Simulation Results

In this section, we evaluate the proposed algorithms by simulations. We consider a cognitive radio network consisting $M = 5$ Rayleigh fading channels. The channel idle probabilities are $\{\theta_m\}_{m=1}^M = \{\frac{2}{3}, \frac{4}{7}, \frac{5}{9}, \frac{1}{2}, \frac{4}{5}\}$. The data rate on a channel m is computed according to the Shannon capacity, i.e., $b_m = \zeta_m \log_2(1 + \frac{P_n h_m}{N_0})$, where ζ_m is the bandwidth of channel m, P_n is the power adopted by users, N_0 is the noise power, and h_m is the channel gain (a realization of a random variable that follows the exponential distribution with the mean $\bar{h}_m$). In the following simulations, we set $\zeta_m = 10\,\text{MHz}$, $N_0 = -100$ dBm, and $P_n = 100\,\text{mW}$. By choosing different mean channel gain $\bar{h}_m$, we have different mean data rates $B_m = E[b_m]$, which equal 15, 70, 90, 20 and 100 Mbps, respectively.

4.6.1 Evolutionary Spectrum Access in Large User Population Case

We first study the evolutionary spectrum mechanism with complete network information in Sect. 4.4 with a large user population. We found that the convergence speed of the evolutionary spectrum access mechanism increases as the strategy adaptation factor α increases (see Fig. 4.3). We set the strategy adaptation factor $\alpha = 0.5$ in the following simulations in order to better demonstrate the evolutionary dynamics. We implement the evolutionary spectrum access mechanism with the number of users $N = 100$ and 200, respectively, in both large and small λ_{max} cases.

4.6.1.1 Large λ_{max} Case

We first consider the case that the number of backoff mini-slots $\lambda_{max} = 100,000$, which is much larger that the number of users N and thus collisions in channel contention rarely occur. This case can be approximated by the asymptotic case $\lambda_{max} = \infty$ in Sect. 4.4.3. The simulation results are shown in Figs. 4.4 and 4.5. From these figures, we see that

- *Fast convergence*: the algorithm takes less than 20 iterations to converge in all cases (see Fig. 4.4).
- *Convergence to ESS*: in both $N = 100$ and 200 cases, the algorithm converges to the ESS $x^* = \left(\frac{\theta_1 B_1}{\sum_{i=1}^M \theta_i B_i}, \ldots, \frac{\theta_M B_M}{\sum_{i=1}^M \theta_i B_i} \right)$ (see Figure the left column of 4.4). At the ESS x^*, each user achieves the same expected payoff $U_n(a_n^*, x^*) = \frac{\sum_{i=1}^M \theta_i B_i}{N}$ (see the right column of Fig. 4.4).
- *Asymptotic stability*: to investigate the stability of the evolutionary spectrum access mechanism, we let a fraction of users play the mutant strategies when the system

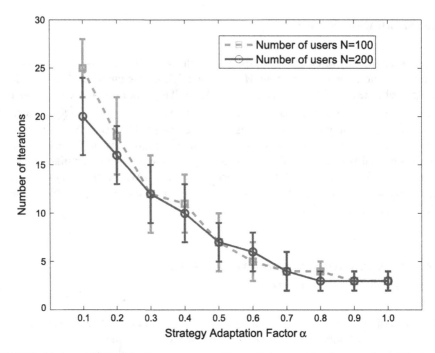

Fig. 4.3 The iterations need for the convergence of the evolutionary spectrum accessing mechanism with different choices of strategy adaptation factor α. The confidence interval is 95 %

is at the ESS x^*. At time slot $t = 30$, $\varepsilon = 0.5$ and 0.9 fraction of users will randomly choose a new channel. The result is shown in Fig. 4.5. We see that the algorithm is capable to recover the ESS x^* quickly after the mutation occurs. This demonstrates that the evolutionary spectrum access mechanism is robust to the perturbations in the network.

4.6.1.2 Small λ_{max} Case

We now consider the case that the number of backoff mini-slots $\lambda_{max} = 20$, which is smaller than the number of users N. In this case, severe collisions in channel contention may occur and hence lead to a reduction in data rates for all users. The results are shown in Figs. 4.6 and 4.7. We see that a small λ_{max} leads to a system performance loss (i.e., $\sum_{n=1}^{N} U_n(a_n(T), x(T)) < \sum_{m=1}^{M} \theta_m B_m$), due to severe collisions in channel contention. However, the evolutionary spectrum access mechanism still quickly converges to the ESS as given in (4.8) such that all users achieve the same expected throughput, and the asymptotic stable property also holds. This verifies the efficiency of the mechanism in the small λ_{max} case.

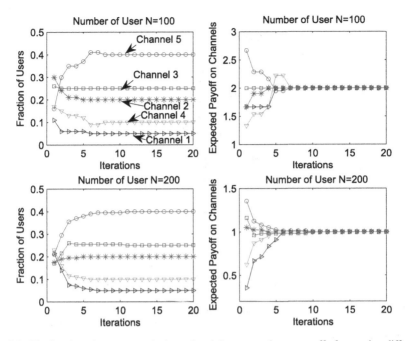

Fig. 4.4 The fraction of users on each channel and the expected user payoff of accessing different channels with the number of users $N = 100$ and 200, respectively, and the number of backoff mini-slots $\lambda_{max} = 100,000$

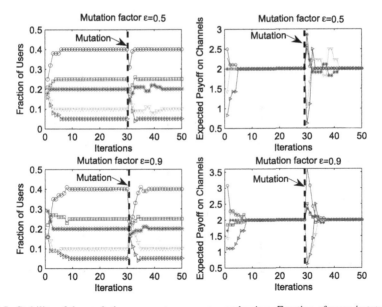

Fig. 4.5 Stability of the evolutionary spectrum access mechanism. Fraction of users in total $N = 200$ users who choose mutant channels randomly at time slot 30 equal to 0.5 and 0.9, respectively, and the number of backoff mini-slots $\lambda_{max} = 100,000$

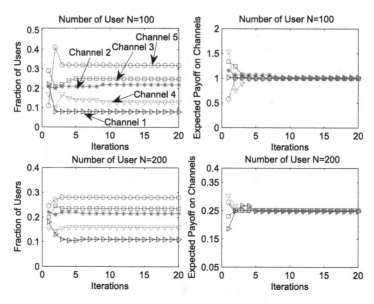

Fig. 4.6 The fraction of users on each channel and the expected user payoff of accessing different channels with the number of users $N = 100$ and 200, respectively, and the number of backoff mini-slots $\lambda_{max} = 20$

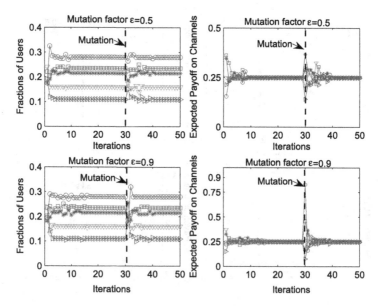

Fig. 4.7 Stability of the evolutionary spectrum access mechanism. Fraction of users in total $N = 200$ users who choose mutant channels randomly at time slot 30 equal to 0.5 and 0.9, respectively, and the number of backoff mini-slots $\lambda_{max} = 20$

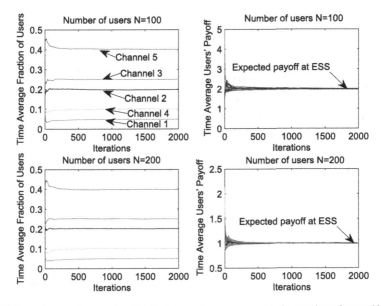

Fig. 4.8 Learning mechanism for distributed spectrum access with the number of users $N = 100$ and 200, respectively, and the number of backoff mini-slots $\lambda_{max} = 100,000$

4.6.2 Distributed Learning Mechanism in Large User Population Case

We next evaluate the learning mechanism for distributed spectrum access with a large user population. We implement the learning mechanism with the number of users $N = 100$ and $N = 200$, respectively, in both large and small λ_{max} cases. We set the memory factor $\gamma = 0.99$ and the length of a decision period $t_{max} = 100$ time slots, which provides a good estimation of the mean data rate. Figures 4.8 and 4.9 show the time average user distribution on the channels converges to the ESS, and the time average user's payoff converges the expected payoff at the ESS. Note that users achieve this result without prior knowledge of the statistics of the channels, and the number of users utilizing each channel keeps changing in the learning scheme.

4.6.3 Evolutionary Spectrum Access and Distributed Learning in Small User Population Case

We then consider the case that the user population N is small. We implement the proposed evolutionary spectrum access mechanism and distributed learning mechanism with the number of users $N = 4$ and the number of backoff mini-slots $\lambda_{max} = 20$. The results are shown in Fig. 4.10. We see that the evolutionary spectrum access

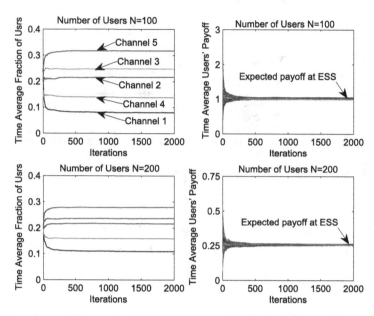

Fig. 4.9 Learning mechanism for distributed spectrum access with the number of users $N = 100$ and 200, respectively, and the number of backoff mini-slots $\lambda_{max} = 20$

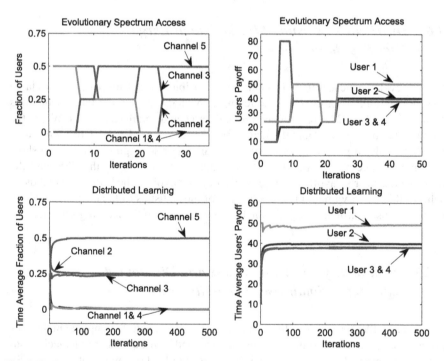

Fig. 4.10 Evolutionary spectrum access and Learning mechanism for distributed spectrum access with the number of users $N = 4$, and the number of backoff mini-slots $\lambda_{max} = 20$

mechanism converges to the equilibrium such that channel 5 has 2 users and both channel 1 and 2 have 1 user. These 4 users achieve the expected throughput equal to 50, 40, 38 and 38 Mbps, respectively, at the equilibrium. It is easy to check that any user unilaterally changes its channel selection at the equilibrium will lead to a loss in throughput, hence the equilibrium is a strict Nash equilibrium. According to [8], any strict Nash equilibrium is also an ESS and hence the convergent equilibrium is an ESS. For the distributed learning mechanism, we see that the mechanism also converges to the same equilibrium on the time average. This verifies that effectiveness of the proposed mechanisms in the small user population case.

4.6.4 Performance Comparison

To benchmark the performance of the proposed mechanisms, we compare them with the following two algorithms:

- *Centralized optimization*: we solve the centralized optimization problem $\max_x \sum_{n=1}^N U_n(a_n, x)$, i.e., find the optimal population state x_{opt} that maximizes the system throughput.
- *Distributed reinforcement learning*: we also implement the distributed algorithm in [2] by generalizing the single-agent reinforcement learning to the multi-agent setting. More specifically, each user n maintains a perception value $P_m^n(T)$ to describe the performance of channel m, and select the channel m with the probability $f_{m,n}(T) = \frac{e^{\nu P_m^n(T)}}{\sum_{m'=1}^M e^{\nu P_{m'}^n(T)}}$ where ν is called the temperature. Once a payoff $U_n(T)$ is received, user n updates the perception value as $P_m^n(T+1) = (1 - \mu_T)P_m^n(T) + \mu_T U_n(T)I_{\{a_n(T)=m\}}$ where μ_T is the smooth factor satisfying $\sum_{T=1}^\infty \mu_T = \infty$ and $\sum_{T=1}^\infty \mu_T^2 < \infty$. As shown in [2], when ν is sufficiently large, the algorithm converges to a stationary point. We hence set $\mu_T = \frac{100}{T}$ and $\nu = 10$ in the simulation, which guarantees the convergence and achieves a good system performance.

Since the proposed learning mechanism in this chapter can converge to the same equilibrium as the evolutionary spectrum access mechanism, we only implement the evolutionary spectrum access mechanism in this experiment. The results are shown in Fig. 4.11. Since the global optimum by centralized optimization and the ESS by evolutionary spectrum access are deterministic, only the confidence interval of the distributed reinforcement learning is shown here. We see that the evolutionary spectrum access mechanism achieves up to 35 % performance improvement over the distributed reinforcement learning algorithm. Compared with the centralized optimization approach, the performance loss of the evolutionary spectrum access mechanism is at most 38 %. When the number of users N is small (e.g., $N \leq 50$), the performance loss can be further reduced to less than 25 %. Note that the solution by the centralized optimization is not incentive compatible, since it is not a Nash equilibrium and user can improve its payoff by changing its channel selection

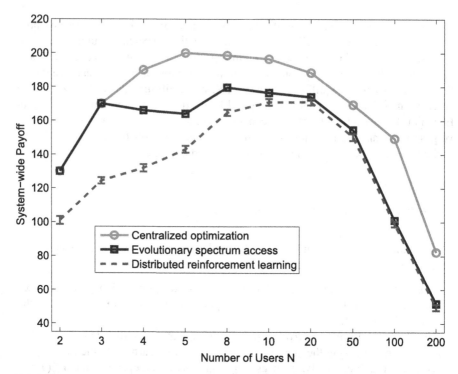

Fig. 4.11 Comparison of the evolutionary spectrum access mechanism with the distributed reinforcement learning and centralized optimization. The confidence interval is 95 %

unilaterally. While the evolutionary spectrum access mechanism achieves an ESS, which is also a (strict) Nash equilibrium and evolutionarily stable. Interestingly, the curve of the evolutionary spectrum access mechanism in Fig. 4.11 achieves a local minimum when the number of users $N = 5$. This can be interpreted by the property of the Nash equilibrium. When the number of users $N = 4$, these four users will utilize the three channels with high data rate (i.e., Channels 2, 3, and 5 in the simulation). When the number of users $N = 5$, the same three channels are utilized at the Nash equilibrium. In this case, there will be a system performance loss due to severer channel contention. However, no user at the equilibrium is willing to switch to another vacant channel, since the remaining vacant channels have low data rates and such a switch will incurs a loss to the user. When the number of users $N = 8$, all given channels are utilized at the Nash equilibrium, and this improves the system performance.

4.7 Summary

In this chapter, we study the problem of distributed spectrum access of multiple time-varying heterogeneous licensed channels, and propose an evolutionary spectrum access mechanism based on evolutionary game theory. We show that the equilibrium of the mechanism is an evolutionarily stable strategy and is globally stable. We further propose a learning mechanism, which requires no information exchange among the users. We show that the learning mechanism converges to the evolutionarily stable strategy on the time average. Numerical results show that the proposed mechanisms can achieve efficient and stable spectrum sharing among the users.

References

1. X. Chen, J. Huang, Evolutionarily stable spectrum access, IEEE Trans. Mob. Comput. **12**(7), 1281–1293 (2013). http://arxiv.org/pdf/1204.2376v1.pdf
2. H. Li, Multi-agent Q-learning for Aloha-like spectrum access in cognitive radio systems, in *IEEE Transport on Vehicle Technology, special issue on Achievements and the Road Ahead: the First Decade of Cognitive Radio* (2009)
3. D. Niyato, E. Hossain, Dynamics of network selection in heterogeneous wireless networks: an evolutionary game approach. IEEE Trans. Veh. Technol. **58**, 2008–2017 (2009)
4. T. Rappaport, *Wireless Communications: Principles and Practice*, vol. 2 (Prentice Hall PTR, Englewood Cliffs, 1996)
5. G. Bianchi, Performance analysis of the IEEE 802.11 distributed coordination function. IEEE J. Sel. Areas Commun. **18**(3), 535–547 (2000)
6. E. Kriminger, H. Latchman, Markov chain model of homeplug CSMA MAC for determining optimal fixed contention window size, in *IEEE International Symposium on Power Line Communications and Its Applications (ISPLC)* (2011)
7. B. Lo, A survey of common control channel design in cognitive radio networks. Phys. Commun. **4**, 26–39 (2011)
8. J.M. Smith, *Evolution and the Theory of Games* (Cambridge University Press, Cambridge, 1982)
9. Q. Zhang, H. Xue, X. Kou, An evolutionary game model of resources-sharing mechanism in p2p networks, in *IITA Workshop* (2007)
10. B. Wang, K.J.R. Liu, T.C. Clancy, Evolutionary game framework for behavior dynamics in cooperative spectrum sensing, in *IEEE GLOBECOM* (2008)
11. H. Kushner, *Approximation and Weak Convergence Methods for Random Processes, with Applications to Stochastic Systems Theory* (The MIT Press, Cambridge, 1984)
12. K.S. Narendra, A. Annaswamy, *Stable Adaptive Systems* (Prentice Hall, Englewood Cliffs, 1989)

Chapter 5
Conclusion

In this book, we propose a novel social cognitive radio networking paradigm, where secondary users share the spectrum collaboratively based on social interactions. The key motivation is to leverage the wisdom of crowds to overcome various challenges due to incomplete network information and limited capability of individual secondary users.

Specifically, we develop three socially inspired distributed spectrum sharing mechanisms: adaptive channel recommendation mechanism, imitation-based social spectrum sharing mechanism, and evolutionarily stable spectrum access mechanism. For adaptive channel recommendation mechanism, inspired by the recommendation system in the e-commerce industry such as Amazon, we treat secondary users as customers and the channels as goods, and secondary users collaboratively recommend "good" channels to each other for achieving more informed spectrum access decisions. For imitative spectrum access mechanism, we leverage a common social phenomenon "imitation" to devise efficient spectrum sharing mechanism, such that secondary users imitate the spectrum access strategies of their elite neighbours to improve the networking performance. For the evolutionarily stable spectrum access mechanism, motivated by the evolution rule observed in many social animal and human interactions, we propose an evolutionary game approach for distributed spectrum access, such that each secondary user evolves its spectrum access decision adaptively over time by comparing its performance with the collective network performance. Numerical results also demonstrate that the proposed socially inspired distributed spectrum sharing mechanisms can achieve superior networking performance.

For the future direction, we can explore other social phenomena such as social reciprocity and leverage the social community structures to design efficient socially inspired distributed spectrum sharing mechanisms. Another important direction is to consider the security issue. How to devise a secure distributed spectrum sharing mechanism against malicious attacks by effectively utilizing the social trust among secondary users will be very interesting and challenging.

© The Author(s) 2015

83

X. Chen and J. Huang, *Social Cognitive Radio Networks*,
SpringerBriefs in Electrical and Computer Engineering,
DOI 10.1007/978-3-319-15215-8_5